INFORMATION SOURCES
FOR RESEARCH AND DEVELOPMENT

Use of
Mathematical Literature

INFORMATION SOURCES
FOR RESEARCH AND DEVELOPMENT

A series under the General Editorship of

R.T. Bottle, B.Sc., Ph.D., F.R.I.C., F.L.A., M.I.Inf.Sc.
and
D.J. Foskett, M.A., F.L.A.

Other titles in the series
Use of Biological Literature (2nd edition)
edited by R.T. Bottle and H.V. Wyatt

Use of Chemical Literature (2nd edition)
edited by R.T. Bottle

Use of Criminology Literature
edited by M. Wright

Use of Earth Sciences Literature
edited by D.N. Wood

Use of Economics Literature
edited by J. Fletcher

Use of Engineering Literature
edited by K.W. Mildren

Use of Management and Business Literature
edited by K.D.C. Vernon

Use of Medical Literature (2nd edition in preparation)
edited by L.T. Morton

Use of Physics Literature
edited by H. Coblans

Use of Reports Literature
edited by C.P. Auger

Use of Social Sciences Literature
edited by N. Roberts

Use of
Mathematical Literature

Editor
A.R. Dorling, B.A., A.L.A.
Formerly Assistant Keeper, British Library, Science Reference Library

BUTTERWORTHS
LONDON–BOSTON
Sydney–Wellington–Durban–Toronto

The Butterworth Group

United Kingdom	Butterworth & Co (Publishers) Ltd
London	88 Kingsway, WC2B 6AB
Australia	Butterworths Pty Ltd
Sydney	586 Pacific Highway, Chatswood, NSW 2067
	Also at Melbourne, Brisbane, Adelaide and Perth
South Africa	Butterworth & Co (South Africa) (Pty) Ltd
Durban	152–154 Gale Street
New Zealand	Butterworths of New Zealand Ltd
Wellington	26–28 Waring Taylor Street, 1
Canada	Butterworth & Co (Canada) Ltd
Toronto	2265 Midland Avenue
	Scarborough, Ontario, M1P 4S1
USA	Butterworth (Publishers) Inc
Boston	19 Cummings Park, Woburn, Mass. 01801

©Butterworth & Co (Publishers) Ltd 1977
First published 1977
ISBN 0 408 70913 8

British Library Cataloguing in Publication Data

Use of mathematical literature.–(Information
sources for research and development).
1. Mathematics–Bibliography
I. Dorling, Alison Rosemary II. Series
016.51   Z6651   77-30014

ISBN 0-408-70913-8

Printed in Great Britain by
The Whitefriars Press Ltd, London and Tonbridge

Preface

Although there are occasions when comprehensive bibliographies of particular subjects are required, situations are far more common in which such an undigested body of information leaves the enquirer feeling no nearer what he sought. If he is in the fortunate situation of having access to someone who knows the field, he may find that a quick word of advice puts him directly on the track of what he wants. He has benefited because he is presented with a selected and evaluated subset of all the information that was potentially available. In the absence of such advice, what steps can he take to find his way about an unfamiliar subject?

The first three chapters of this book are concerned with the character of mathematical literature and the ways in which it can be effectively approached. Chapters 2 and 3 are mainly descriptive of particular tools; Chapter 1 is a discussion of their uses. No mathematical background is required for these chapters.

In the remaining chapters a group of specialists give critical accounts of the major literature in their fields. The subjects are presented in the order in which they would be found in the American Mathematical Society's subject classification to *Mathematical Reviews* (except that probability has been brought forward to take its place with measure theory). A glance at the index will show how much more complex are the connections between subjects than any linear arrangement of chapters can reflect.

It is hoped, then, that this book will be of value to students embarking on research in mathematics; to working mathematicians, whether academic, commercial or industrial, exploring fields other than their own; and to all those, such as teachers, librarians, translators, who have to handle mathematical literature or are interested in the structure of mathematics.

I should like to put on record my gratitude to my contributors for all their work, and my thanks to too many people to be individually named for help ranging from long discussions on the aim and content of this book to the odd word which has sparked off a train of thought.

A.R.D.

Contributors

J. Frank Adams, M.A., Ph.D., F.R.S., Lowndean Professor of Astronomy and Geometry, Trinity College, Cambridge

Norman L. Biggs, B.A., Royal Holloway College, University of London

Susan Brookes, B.Sc., Diploma of the New Zealand Library School, formerly British Library, Science Reference Library, London

James M. Brown, M.A., M.Sc., The University of Manchester Institute of Science and Technology

Paul M. Cohn, M.A., Ph.D., Professor of Mathematics, Bedford College, University of London

Alison R. Dorling, B.A., A.L.A., formerly British Library, Science Reference Library, London

Ivor Grattan-Guinness, M.A., M.Sc., Ph.D., Middlesex Polytechnic, Enfield

Wilfrid A. Hodges, M.A., D.Phil., Bedford College, University of London

A. Geoffrey Howson, M.Sc., Ph.D., F.I.M.A., University of Southampton

Rhys Price Jones, M.A., M.Sc., Royal Holloway College, University of London

Geoffrey T. Kneebone, M.Sc., Ph.D., Reader in the Foundations of Mathematics, Bedford College, University of London

Peter McMullen, M.A., Ph.D., University College London, University of London

Peter M. Neumann, M.A., D.Phil., The Queen's College, Oxford

Alan R. Pears, M.A., Ph.D., Queen Elizabeth College, University of London

Ian N. Sneddon, O.B.E., M.A., D.Sc., Simson Professor of Mathematics, University of Glasgow

S. James Taylor, B.Sc., Ph.D., Professor of Pure Mathematics, University of Liverpool

Contents

1

The Role of the Literature in Mathematics

A.R. Dorling

1.1 INTRODUCTION

There was a story current in Oxford at one time that when one of the colleges was having a new building in which a mathematics don was to have a room, he asked for a sand-pit. This occasioned some surprise, but he explained that he found it the best way of working, because things could be altered so easily. Even more freely than on a blackboard, an idea could be suggested, manipulated, conveyed to others. The story may have been apocryphal, but it highlights by contrast the static nature of the printed word, and suggests one of its functions: to record the finally worked-out idea. But no piece of writing is purely archival in character. The paper which establishes one's priority in reaching a particular result also makes that result generally available to be incorporated in future developments. Comparison with the sand-pit suggests that it may not be the most effective way of communicating the result, but the fact that it is in print implies both that the intuitively conveyed idea has been established on a sounder footing and that it can potentially reach a much wider audience than that to which it can be conveyed by gesture and word-of-mouth.

There has been relatively little work reported which examines the effectiveness of mathematical literature as a means of communication. The International Mathematical Union (cf. Section 2.2) has been concerned with the problem and has encouraged the exchange of reviews between reviewing journals so that a broader picture of what is happening can be obtained from a single source. The Society for Industrial and Applied Mathematics organised a symposium which was reported in I.E. Block and R.F. Drenick (eds.), 'A symposium on the publication of mathematical literature' (*SIAM Review,* **6**, 431-454, 1964). Since then the pattern of available literature has been substantially altered—for example, by the American Mathematical Society's provision of a current awareness publication and of greatly increased subject access to *Mathematical Reviews* (cf. Section 3.3). The growth in mathematical literature is discussed in A.J. Lohwater, 'Mathematical literature' (*Library Trends,* **15**, 852-867, 1967).

Techniques have been developed for investigating aspects of publication practice, and the term 'bibliometrics' has been coined to cover them. In the context of this book, the most interesting are those associated with the Bradford–Zipf distribution. This arose from observations on the scatter of papers cited in bibliographies among the cited journals. The statistical study of the resulting distributions makes it possible to predict the size such a bibliography would have if complete, and more usefully enables one to say what proportion of the papers is covered by the n most productive journals. It is the concentration effect described by the Bradford–Zipf distribution which makes it reasonable for the Mathematical Offprint Service (cf. Section 2.3) to have covered only 60 journals and for *Science Citation Index* (cf. Section 3.3), although concerned with the whole of science and various fringe areas, to cover fewer than 3000 journals, a small proportion of the total published. The same effect operates in relation to the Russian mathematical journals which are translated, a matter which must afford some comfort to the non-Russian-speaking mathematician. A quick examination of the bibliographies to the chapters in this book will suggest that a similar relationship exists between books and publishers as between papers and journals, and that if one were to scan the publisher's details of new books from quite a small number of publishers, one would keep informed of the bulk of relevant new material. A review and bibliography is given by R.A. Fairthorne, 'Empirical hyperbolic distributions (Bradford–Zipf–Mandelbrot) for bibliometric description and prediction' (*Journal of Documentation,* **25**, 319-343, 1969). A possible explanation of the mechanism behind the distribution (in a much wider context than a bibliometric one) was suggested by G. Scarrott, 'Will Zipf join Gauss?' (*New Scientist,* 16 May, 402-404, 1974).

It should be added that Chapter 2 does not attempt to identify a core group of major mathematical journals. Increasingly with the publication of specialist journals (cf. Section 2.7), that group will depend upon the subject area in which one is working. It will also be something which changes with time.

1.2 THE AUTHOR APPROACH TO THE LITERATURE

The relatively recent (in terms of the lifetime of *Mathematical Reviews*) introduction of increased subject access mentioned in the introduction may have arisen because the provision of a subject index was not feasible before computerisation (cf. Section 2.3); but it suggests rather that there did not exist a sufficiently strongly felt need for such a tool. To find out what the latest results in a particular field were, one kept track of the papers of a small group of people working in that field. Again the reasonableness of such an approach may be related to the Bradford–Zipf effect: one does not attempt to identify or keep track of everything which is going on in the field, only the 'important' things. The publication of *Science Citation Index* (cf. Section 3.3) has facilitated the identification of who is working on particular problems by identifying those authors whose bibliographies include a known relevant paper. By carrying out a 'snowballing' search and finding which authors have cited these authors, a picture of the work being done can be built up. The possibility of closure of the set of authors is of theoretical rather than practical interest.

The *Science Citation Index* approach is an interesting one, but it can be tedious, and it does require a starting point. Bibliographies covering limited subject areas are mentioned in Section 3.4, and of course the volumes of collections of reviews from *Mathematical Reviews* in particular subject areas (cf. Section 2.3; References 12, 77, Chapter 9; Section 13.1) are valuable examples. However, unless a bibliography has a very restricted time span, it is not the most useful source for identifying people currently working in the field. Far more valuable, if one can be found, is a recent conference on the subject, and these are discussed in Section 1.3. Alternatively, if one knows from a bibliography that a particular mathematician has contributed substantially to a field, the possibility arises of a 'Festschrift' to him of research papers from people active in the areas to which he has contributed (cf. Reference 106, Chapter 12).

While knowledge of appropriate authors may enable one to replace a subject search by an author search, the results of the two will not be the same. The appearance of work by the same author on differing (at least as far as the arrangement of the library catalogue, abstracting journal or bibliography you are consulting goes) subjects suggests relationships between those subjects which are not apparent unless one is familiar with the field. The author section of the index to this book can be used in this way to broaden the picture given by any particular chapter. *Referativnyi Zhurnal. Matematika* (cf. Section 3.3) encourages an author approach to a subject search by following the name in the author index by a number leading into the appropriate section of the subject index. As Section 3.3 makes clear, *Referativnyi Zhurnal* is a tool amenable to use by non-Russian-speakers. It is compared with *Mathematical Reviews* in A.J. Lohwater's paper mentioned on p. 2, but one would like to see a more recent study.

1.3 CONFERENCES

A conference (congress, symposium, etc.) seen from the point of view of a delegate is a very different thing from the same conference seen by a reader of the published proceedings, a point which is emphasised by the following extract from the introduction to *Intuitionism and proof theory* (edited by A. Kino, J. Myhill and R.E. Vesley), North-Holland, Amsterdam (1970) (cf. Reference 29, Chapter 6), which is the proceedings of a conference held in Buffalo, New York, in 1968:

> There are some discrepancies between the list of papers contained in this book and those actually read at the Conference. In particular, Takeuti gave a course of lectures on proof theory, and Troelsta ten on intuitionism; the Takeuti lectures will be published in an extended form by North-Holland and Troelstra's have already been published as Vol. 95 in the Springer lecture-note series ('Principles of Intuitionism'). A paper by Gilmore . . . was withdrawn because the system presented turned out to be inconsistent, and one by Sabbagh . . . because its results were found at the last minute to be properly included in those of McKay, *J. Symb. Logic* 33 (1968), 258-264. Papers by Krasner . . . and De Jongh . . . were not received in time for inclusion in this volume. A paper by Hull . . . had already been accepted for publication in the *Zeitschrift für Mathematische Logik* before the conference was held (*Z. Math. Logik* 15 (1969), 241-246). Kleene's paper . . . was not included because of its length; it has subsequently appeared as number 89 of the *Memoirs of the American Mathematical Society* (1969). The paper XX of Feferman contains only a part of the material presented at the conference. The rest is contained in a privately circulated manuscript . . . [quoted by courtesy of the North-Holland Publishing Company Limited] .

Both the delegate and the reader benefit by identifying who is contributing to a particular field, and for the delegate the informal part of the conference, outside the formal lecture sessions, may be of greater value than the lectures, at least a proportion of which may not be intelligible to him at first hearing. The quotation confirms that the printed source is the place to go for established results. It also suggests some of the problems which arise in producing the proceedings of a conference and which make conference material more difficult as a source than articles in journals. The proceedings have to be published as a unit at a particular time, whereas the combination of articles in one issue of a journal is as a rule fortuitous. The result of this is that the proceedings are likely to be less current than the same material produced as separate articles.

Again, proceedings in book form are less sure of being abstracted than is material in journals, a point which was made in the report of a Unesco Ad-hoc Sub-committee ('The flow of information from international scientific meetings', *Unesco Bulletin for Libraries,* **24,** 88-97, 1970), which encouraged the production of proceedings as special issues of journals. A solution which is half-way between this and publication of an isolated volume is the appearance of the proceedings in a well-established monograph series. 'Lecture Notes in Mathematics' (cf. Section 2.9) has provided a home for the proceedings of many conferences which would quite possibly not otherwise have seen the light of day, and editors of the series lay down quite stringent guidelines to ensure the completeness of the material.

The Unesco Ad-hoc Sub-committee report raises a number of other points of interest. One is the need for papers to be refereed before they are published, not simply included because they were presented at the conference. Where papers are invited this may have been done (particularly if there were conference preprints—duplicated material produced for delegates; cf. Section 1.4); the problem arises with contributed papers. Where published proceedings do not make it clear whether account has been taken of discussion which arose at the conference or whether the material has been subjected to any kind of refereeing process, then results have to be looked at more critically than would otherwise be the case. One of the requirements of 'Lecture Notes in Mathematics' is that abstracts are not accepted—the full derivation of results must be given.

Another recommendation of the Unesco Ad-hoc Sub-committee report is that international conferences could perform a valuable service in commissioning reviews of subjects. The educational function which this suggests is often undertaken in a different way by attaching a seminar or lecture series to a conference (as was done in the conference on intuitionism and proof theory). This has the advantage from the point of view of the conference organisers that it may make it easier to obtain grants to support the conference.

In the case of the conference on intuitionism and proof theory, the introduction tells one where the material not included in the book is to be found, but publication of conference proceedings distributed in more than one source or appearing in more than one form (for example, as a group of articles in a journal and as a separately available book) is by no means unknown and suggests the need to give the fullest possible information in any reference to a paper in a conference. An example from the references in this book is given by the conference on combinatorics edited by M. Hall, Jr., and J.H. van Lint (cf. References 29, 38, 94 and 101, Chapter 7; Reference 37, Chapter 9). *Mathematical Centre Tracts,* **55–57** (1974) contains the papers given in the first three of the six sections of the conference, and does not mention the appearance of proceedings in book form.

1.4   UNPUBLISHED MATERIAL

Conference material is useful as a compact source for obtaining a picture of what is going on in a particular field and who is working in it, but the published proceedings may already give an out-of-date picture. Where currency is important, unpublished material becomes of value. It falls into two categories, although the borderline between them can sometimes be narrow. The first is material of a research nature. This will include draft papers for comment, which will of course only be available to a very restricted group, and conference preprints made available to delegates. In the latter case, the material may become generally available in the form of a journal article or in published proceedings. However, if one knows that a conference is to take place in which one is interested but which one cannot attend, it is possible to write to the organisers to find out whether published proceedings are planned. Particularly where it is not intended that proceedings will be published, this may enable one to obtain preprints. Unpublished research material may also result from seminars given at universities or research institutes.

It is this latter type of material which comes close to the second category: lecture notes of expository character produced by mathematics departments or institutes to meet the needs of their students. These will tend to be intermediate between monographs and journals articles, or will fill in areas where no suitable monograph is available. As a rule, such material would be made available on request. The difficulty is to get to know of its existence. Up until volume **6(5)** of *Contents of Contemporary Mathematical Journals* (cf. Section 3.3), the American Mathematical Society listed newly available material of this kind from a number of principally American or Canadian universities—for example, Carleton Lecture Notes and the Carleton Mathematical Series (cf. References 9 and 64, Chapter 8). In the classified arrangement which replaced the journal-by-journal listing of contents and which continued in *Current Mathematical Publications,* it is no longer possible to search specifically for this type of material. The same type of material attached to a conference is often not published in the conference proceedings, although one would hope that editors would, in general, follow the example of those of *Intuitionism and proof theory* (cf. Section 1.3) and indicate where it is to be found.

It may be easier to establish the existence of unpublished material from a more advanced lecture series or seminar, from the knowledge that a particular mathematician is visiting a department or institute, or that that organisation regularly holds a research seminar or lecture series. There are in particular a number of important French seminars of this kind: the Séminaire Bourbaki now appears in 'Lecture Notes in Mathematics' (cf. Section 2.9). Others, such as the Séminaire of the Institut de Mathématiques of Strasbourg University (referred to at the end of Chapter 10 and in Reference 84, Chapter 11), the Séminaire Dubreil from the Institut Henri Poincaré (References 126 and 143, Chapter 8) and the Séminaire Cartan from the École Normale Supérieure (References 75, 80 and 81, Chapter 11), would have to be obtained from their originating organisations.

I prefer to exclude theses from the category of unpublished material, since they are published in some countries and worthwhile material from a thesis is likely to find its way into print. Where completeness of search is required—for example, to avoid duplication of a research topic—tools such as *Dissertation Abstracts* (University Microfilms, Ann Arbor) and the *Index to Theses Accepted for Higher Degrees in the Universities of Great Britain and Ireland* (Aslib, London) can be used.

1.5 BOOKS

The emphasis so far in this chapter has been on keeping in touch with recent research results. But of course this is not the only use to which mathematical literature is put. Mention has already been made of the important expository material to be found among the unpublished notes from universities. In general, however, if one is looking for material of this character, one would expect to find it in book form. It has already been suggested in the introduction to this chapter that one way of keeping track of what is being published is to scan the publisher's details from a small group of publishers. In most cases these details will give a summary of contents for each book, so that the level of the book can be determined reasonably well. Now that *Current Mathematical Publications* has a classified arrangement (cf. Section 1.4), it is no longer possible to scan it just for books, as could be done with 'Recent book publications' in *Contents of Contemporary Mathematical Journals* and *New Publications,* which preceded it. A synoptic view of book publication is probably best obtained from *Internationale Mathematische Nachrichten* (cf. Section 2.2), which has a section listing new books country by country, followed by a review section (the reviews are in German).

A particular category of book which merits some mention is collected papers of mathematicians. As with the conference proceedings which appear in more than one form (cf. Section 1.3), these can give rise to references which are difficult to trace; libraries are reluctant to hold the same material more than once. Where a library does not hold journals published before a certain year, a reference to a paper can be pursued by seeing whether the collected works of its author have been published. The converse situation, where the reference is to the collected works and does not indicate the original source of the paper, may be very difficult indeed to resolve. Alternative sources for bibliographies of the work of particular mathematicians are 'Festschriften' for living mathematicians and obituary notices such as those in the *London Mathematical Society. Bulletin* or, for Russian mathematicians, *Russian Mathematical Surveys* (cf. Section 2.4). At a later stage, a biography (cf. Section 3.5) should also provide such a bibliography.

1.6   FUTURE TRENDS

The particular literature we use is all too often the result of chance: a colleague points out an article in a journal we have not been in the habit of scanning, and we find there are also other things to interest us there; or a publisher's salesman draws our attention to an important new book, and we request our library to buy it without having seen it or waited for a review or checked what else has recently been published in the same field. Where the use does not arise from chance, it may arise from long-established habit. The wider our knowledge of the use of libraries and information services is, the better those habits are likely to be, but the pattern of the literature is a changing one, and to make effective use of material, it is essential to be alert to changes—it is also of interest in its own right! Two examples mentioned in the following chapter are the change away from the refereeing system towards communication via editors, and the development of journals specialising in restricted fields (cf. Section 2.7). A letter from J.B. Conway in the *American Mathematical Society. Notices,* **23**(1) (January 1976) pointed out that as a referee he was not usually informed whether his recommendation was followed, let alone being sent an offprint of the article—scarcely an encouragement to mathematicians to give of their time in this way.

It is obviously changes in the field of journal publishing which are of major concern. If economic stringency causes libraries to cut the number of their journal subscriptions, journals whose subject coverage is limited are likely to be more affected than those with broad coverage, simply because their potential audience is smaller. The personal subscriber may have to weigh the value to him of a specialist journal against the value of society membership. As a society member he will receive one or more of the publications of that society as part of his subscription or at a rate considerably below that of many commercially produced journals. Factors like these will in turn affect the choice of journals to which a paper will be submitted. Although it will doubtless remain true that there is a core of journals which cover a substantial part of the work in any particular subject, the membership of that core will change.

2

Major Organisations and Journals

A.R. Dorling

2.1 INTRODUCTION

As the primary printed source of information for a working mathematician must be the journal article, it is the purpose of this chapter to consider a number of important mathematical organisations in their roles as producers of literature, and review some general features of periodical publishing in mathematics.

Before doing so, it is worth considering the tools which can be used for tracing the existence of periodicals. (Tools for locating particular articles are described in Sections 3.3 and 3.4.) References to journal articles frequently give the journal title in abbreviated form (full titles have been given in references throughout this book), and a number of different systems of abbreviation are in use. In future, abbreviations should be governed by two standards from the International Standards Organisation:

ISO   4, *Documentation—International Code for the abbreviation of titles of periodicals (1972)*

ISO 833, *Documentation—International list of periodical title word abbreviations (1974)*

The list is kept up to date by six-monthly supplements issued by the International Serials Data System (ISDS) centre in Paris.

The single most useful source for expanding an unfamiliar abbreviation is the index issue of *Mathematical Reviews,* which each year gives a list of source journals for the articles abstracted that year. The list is in alphabetical order of abbreviations, giving the full title and place of publication (the latter being important to avoid confusion between journals with similar titles). The proceedings, journals, etc., of societies, institutes and similar bodies are not listed under the body: *London Mathematical Society. Proceedings* appears as *Proc. London Math. Soc.* While differences in abbreviations should not, in general, make an enormous difference to the order of titles, problems may occur with different systems of transliteration from non-Roman scripts. Here *Mathematical Reviews* helps by following the journals list by a comparative table of the transliterations of Cyrillic used by themselves (they adopted the system used by the *Zentralblatt für Mathematik und ihre Grenzgebiete*), *Bulletin Signalétique, Applied Mechanics Reviews, Science Abstracts,* the Library of Congress and the *Journal of Symbolic Logic.*

Where further information about a journal is needed, the source giving the most is *Ulrich's International Periodicals Directory, 16th edn 1975-6,* Bowker, New York (1975). This is a classified list under broad subject headings, with an alphabetical title index. The mathematics section is not large, but references to relevant journals in other sections are included. The information covered includes International Standard Serial Number (a unique eight-digit identifying number allocated by the ISDS centre), translation of the title for foreign titles, language(s) of the text, year first published, sponsoring body, publisher's name and address, editor and the abstracting services which cover the journal.

2.2   INTERNATIONAL MATHEMATICAL UNION

There was a growing feeling during the last decade of the nineteenth century that some form of international mathematical assembly was needed. In 1896 the mathematicians of the Eidgenössische Technische Hochschule in Zürich formed a committee to organise such an assembly and issued invitations in January of the following year for an assembly the following August in Zürich. At this assembly the title of 'International Congress of Mathematicians' was adopted, and the Société Mathématique de France agreed to organise a subsequent congress. This took place in 1900, and thereafter the congress took place every four years, with a break from 1912 until 1920 (when a new series of congresses was begun), and a further break from 1936 to 1950.

Prior to the 1950 congress, delegates from 22 of the countries involved met and drafted a constitution for an 'International

Mathematical Union', which was presented to the congress and given its formal approval. The intention of the IMU was to promote international co-operation by supporting the International Congress of Mathematicians and other international activities likely to aid the development of pure and applied mathematics, or promote mathematical education. It came into existence officially in September 1951, by which time ten countries had adhered (the United Kingdom through the Royal Society). The first general assembly was held in 1952 with financial aid from Unesco, and in the same year the IMU became federated to the International Council of Scientific Unions (ICSU), a body receiving financial aid from Unesco and having consultative status vis-à-vis various United Nations organisations such as the International Atomic Energy Agency and the World Health Organization.

One of the concerns of the 1952 general assembly of the IMU was to establish an international mathematical news bulletin. The means of doing so were available thanks to the Österreichische Mathematische Gesellschaft. The *Nachrichten der Mathematischen Gesellschaft in Wien,* founded in 1947 and retitled *Nachrichten der Österreichischen Mathematischen Gesellschaft* in 1949, not only contained the news of meetings of the Society, abstracts of books and articles by members, etc., which one would expect in the news publication of a national mathematical society; it also contained news of congresses, societies and university mathematicians all over the world. From 1951, this section was entitled *Internationale Mathematische Nachrichten: International Mathematical News: Nouvelles Mathématiques Internationales.* The 1952 assembly agreed to negotiate a contract with the Österreichische Mathematische Gesellschaft to continue its publication in a way suitable to the needs of the IMU. The three-language title became the main title of the journal from 1962, and thereafter IMU news has been published in English or French, the official languages of the IMU, with the remaining international news and news of the Austrian Society in German.

In the case of the International Congress of Mathematicians (ICM) also, the IMU does not have the resources to act as its own publisher, but has the responsibility for finding a publisher. The *Proceedings* of successive congresses consequently have different publishers and have appeared at varying intervals after the congresses have taken place. The result is that some libraries will treat *International Congress of Mathematicians. Proceedings* as a journal (and references to the *Proceedings* in this book are made as though to a journal), either under its English title, or in the corresponding French form: *Congrès International des Mathématiciens. Actes*; others will treat each as an individual book. To aid in tracing the *Proceedings* in this form, details of the post-world war congresses follow:

1950, Harvard, 2 vols, American Mathematical Society, Providence, R.I. (1952)

1954, Amsterdam, 3 vols, Noordhoff, Groningen; North-Holland, Amsterdam (1954-1957)

1958, Edinburgh, Cambridge University Press, Cambridge (1960)

1962, Stockholm, Institut Mittag-Leffler, Djursholm (1963)

1966, Moscow, Mir, Moscow (1968)

1970, Nice, 4 vols, Gauthier-Villars, Paris (1971)

1974, Vancouver, 2 vols, Canadian Mathematical Congress, Montreal (1975)

Although the existence of the IMU will only impinge on most mathematicians' consciousness through the ICM, consideration of some of its other activities shows ways in which they may have been indirectly helped by it. The IMU has established a number of commissions and committees. The International Commission on Mathematical Instruction (ICMI) and the earlier International Commission on the Teaching of Mathematics, established at the 1908 Congress, are mentioned in Sections 4.3 and 4.4. ICMI has a semi-independent status with its own secretariat and own journal, *L'Enseignement Mathématique.* Another committee has responsibility for the *World Directory of Mathematicians* (cf. Section 3.5). Publication of this was made possible with the co-operation of the Tata Institute of Fundamental Research in Bombay, with a first edition in 1958 and a second in 1961. The third edition was produced in time for the ICM Moscow meeting in 1966, and the project continued with a fourth edition in 1970, produced in Sweden.

A commission to produce a directory of mathematical symbols was disbanded in 1954, and the matter subsequently came under the aegis of the Commission on Scientific Publications, as did the work of the Commission on Abstracting and Reviewing, which was responsible in 1953 for arranging for the exchange of reviews between *Mathematical Reviews* and the *Zentralblatt für Mathematik und ihre Grenzgebiete.* The rationalisation of abstracting services in particular subject areas is the concern of the ICSU Abstracting Board, set up in 1953, and although the early work of the IMU in this field was done independently of ICSU-AB, it has maintained its links with ICSU.

Apart from the question of mathematical symbols, which has now become the concern of the International Standards Organization (ISO/TC46/SC4/WG1), which is preparing a standard primarily with the needs of computerised handling of the bibliographical information in mind, the Commission on Scientific Publications considered the possibility of an annual volume of expository articles, and found through its contacts with commercial publishers that Academic Press were interested in such a venture independently of any financial support from the IMU. *Advances in Mathematics* was first published in 1961.

2.3 AMERICAN MATHEMATICAL SOCIETY

The AMS is one of the most important, if not the most important, publishers of mathematical literature. Its longest-standing periodical, the *Bulletin,* was founded in 1894 and combines the function of official organ of the society, giving notices of meetings, obituaries, etc., with the publication of research articles. A home for longer articles was set up with the foundation of the *Transactions* of the society in 1900. *Mathematical Reviews* (cf. Section 3.3) was founded in 1940 and, as we have seen in the last section, was becoming more international in coverage in the 1950s. In 1950 two further periodicals were founded: the *Proceedings* took over the shorter research articles which had previously been published in even-numbered issues of the *Bulletin*; the *Memoirs* contains papers of such length as almost to count as research monographs. One further periodical was started after the AMS general office moved from New York to Providence, R.I., in 1951, the *Notices* in 1953. This took over most of the announcement functions of the *Bulletin,* becoming the news journal of the society and leaving the *Bulletin* to act as the place for quick short announcements of research results. The *Notices* has a section of abstracts which performs a similar function. A feature of the *Notices* of interest to anyone submitting articles for publication is the section in the February and August issues containing information on the claimed and observed lags between acceptance and publication for a selection of mathematical research journals.

In the 1960s the AMS decided to computerise the production of *Mathematical Reviews.* Whereas a major consideration behind the computerisation of, for example, *Chemical Abstracts* and *Index Medicus* was the need to cope with a situation in which the production of indexes was swallowing an ever-increasing proportion of the budget and taking longer and longer as the services grew, this situation was not really operative for the AMS: *Mathematical Reviews* at this time had only an author index. Rather, computerisation would enable AMS to produce a number of other services which would not be feasible by manual means. The most ambitious service planned, with a supporting grant from the National Science Foundation, was the Mathematical Offprint Service (MOS), planned to become operative in 1968. The subject headings in use in *Mathematical Reviews* were enhanced to provide a detailed classification, first published as an appendix to

Mathematical Reviews, **39** (1970). Using terms from the AMS (MOS) classification linked with authors' names to form Boolean equations, with the possibility of excluding unwanted languages, the subscriber to the MOS service could have a profile of his interests matched against the articles published in 60 core mathematical journals. He would then receive offprints of articles achieving a match with his profile and a title-listing of articles of secondary interest. Already in 1968 M. Cooper was criticising the AMS in her article 'Current information dissemination: ideas and practices' (*Journal of Chemical Documentation,* **8**, 209, 1968), both for intending to go straight into operation without a pilot scheme first, and for failing to survey the potential market to see whether there was a need for such a service. The service has not continued, but a more immediate reason for that was the copyright problems which the provision of offprints entailed. On the positive side, however, can be placed *Current mathematical publications* and *Index of mathematical papers* (cf. Section 3.3). The conversion to machine-readable form has also made possible the publication of collections of reviews from *Mathematical Reviews* in particular subject areas. One of the most recent is W.J. LeVeque (ed.), *Reviews in Number Theory as Printed in Mathematical Reviews, 1940-1972,* 6 vols (1974).

In the field of conferences, the AMS established two important serials: *Proceedings of Symposia in Applied Mathematics* (1950-1967), which continued as *SIAM-AMS Proceedings,* produced jointly with the Society for Industrial and Applied Mathematics (cf. Section 14.4 for other journals from the same source); and *Proceedings of Symposia in Pure Mathematics* (1959–   ). These are the proceedings of AMS summer research institutes, with a few volumes devoted to symposia held in conjunction with AMS sectional meetings. The AMS also publishes the *Conference Board of the Mathematical Sciences. Regional Conference Series in Mathematics* (1970–   ), each number of which gives the expository lectures from a conference sponsored by the CBMS (see, for example, Reference 93, Chapter 9).

It remains to consider the AMS's involvement in two other important areas of mathematical literature: book production, and translation. The series of 'Colloquium Publications' was started in 1905 and rather belies its title. While earlier volumes such as Osgood's *Topics in the theory of functions of several complex variables* (cf. Reference 69, Chapter 11) may have been based on a lecture series, it is difficult to believe that this is true of all the volumes in the series. At most, a didactic intention can be attributed to them which would be lacking in a straightforward monograph series. The 'Mathematical Surveys' series was started in 1943, with J.A. Shohat and J.D. Tamarkin, *The problem of moments,* and was obviously intended to provide bridging material between the textbook and research papers in areas which were not served by monographs. Volumes in the series have not appeared at a great rate: J.R. Isbell's *Uniform Spaces,* Mathematical Surveys 12 (1964), was followed by C. Pearcy's *Topics in operator theory,* Mathematical Surveys 13 (1974). This is presumably an instance of the perennial difficulty of persuading creative mathematicians to divert time to the writing of survey volumes.

The need to gain access of material in unfamiliar languages has led the AMS to publish mathematical dictionaries for Russian and Chinese (cf. Section 3.6). It also started a periodical *Translations* in 1949 to contain important material drawn from a wide range of sources, symposia, festschriften, etc. In this it differs from most other translation periodicals, which are cover-to-cover translations or selections from particular journals. The AMS also publishes the latter type of serial. *Chinese Mathematics* is a translation of *Acta Mathematica Sinica* (1951–   ) from 1962. The AMS monograph series 'Translations of Mathematical Monographs' contains a number of translations of important Russian works—for example, Fuks' *Introduction to the theory of analytic functions of several complex variables* (cf. Reference 85, Chapter 11). Some Chinese material is also included—for example, L.K. Hua, *Additive theory of prime numbers,* Translations of Mathematical Monographs 13 (1965)—and the series now (1976) numbers 46 volumes.

2.4   LONDON MATHEMATICAL SOCIETY

The London Mathematical Society is not a publisher on the scale of the American Mathematical Society, but its activities as a producer of literature can usefully be compared.

It was founded in 1865 (a history of the society can be found in E.F. Collingwood, 'A century of the London Mathematical Society', *London Mathematical Society. Journal,* **41**, 577-594, 1966) and started to publish its *Proceedings* in the same year. A useful survey of the major publications which appeared in the *Proceedings* up to the 1920s is given in H. Davenport, 'Looking back' (*London Mathematical Society. Journal,* **41**, 1-10, 1966). The *Journal* was started in 1926 to take shorter papers and such things as obituaries. The *Bulletin* (1969–   ) now acts as the place for quick publication of research results, as well as containing survey articles of particular areas of mathematics. It was originally hoped to have one of these in each issue, but it has proved impossible to find enough contributors for this. The *Bulletin* also acts as the news journal for the Society.

In the field of monograph series, the LMS has combined with commercial publishers to produce two series. The first, the 'London

Mathematical Society Lecture Note Series' published by the Cambridge University Press, contains material of postgraduate level. The second is 'London Mathematical Society Monographs', published by Academic Press, which started in 1970 with Wall's *Surgery on compact manifolds* (cf. Section 13.3).

The LMS also turns to commercial publishers for the conferences which it sponsors (see, for example, Reference 30, Chapter 9), and does not have a periodical conference publication.

In the field of translation, the LMS publishes *Russian Mathematical Surveys,* (15, 1960–   ), which covers survey articles and biographical material from *Uspekhi Matematicheskikh Nauk.* This translation is now sponsored jointly with the British Library Lending Division.

2.5 OTHER NATIONAL SOCIETIES

The publication activities of the LMS are more typical of what one would expect of a national mathematical society than are those of the AMS: one or more journals providing news; a place for articles of original research and others surveying the present state of a narrow field or historically surveying the work of an individual or school; and possibly a place for quick announcement of research results which may later be written up in more detail elsewhere. The pattern in any one country will depend on the relationship between purely mathematical societies and national societies. For example, Moscow has one of the oldest mathematical societies, the Moskovskoe Matematicheskoe Obschchestvo, founded in 1864, a year before the LMS, and publishing its *Trudy* (1952–   ) and *Matematicheskii Sbornik* (Novaya seriya, 1936–   ). But it is the *Doklady* (cf. Section 2.8) of the Akademiya Nauk SSSR which has a section for results with abstracts of proofs comparable with the abstracts section of *American Mathematical Society. Notices* (cf. Section 2.3). Similarly, in France it is not the *Bulletin* of the Société Mathématique de France to which mathematicians turn for first publication of their results, but to the *Académie des Sciences. Comptes Rendus Hebdomadaires des Séances. Série A. Sciences mathématiques.* This situation may be a fluid one. In the past both the *National Academy of Sciences. Proceedings* in the United States and the *Royal Society. Proceedings* in the United Kingdom have held a more important role for the mathematical community than they do now.

The news function will probably extend beyond the society to other activities—for example, conferences—in the country. In the case of the Österreichische Mathematische Gesellschaft, we have seen in

Section 2.2 how they achieved an international news function. The society's periodicals are unlikely to be restricted to articles from members of the society or even nationals of the country. The *Commentarii Mathematici Helvetici* (1929–   ), published by Birkhäuser for the Schweizerische Mathematische Gesellschaft, has long had an international status, and accepts articles in any one of French, German or English. The proceedings of conferences sponsored by the society, if published at all, are likely to be published by a commercial publisher.

2.6   UNIVERSITY PUBLICATIONS

I am not concerned in this section with journals published by the established university presses such as the Princeton University Press or the Cambridge University Press. Although their origin in a university will predispose them to publish material of a certain academic level, their role is strictly comparable with that of other commercial publishers. Rather I am thinking of journals originating from a mathematical department or institute which have achieved international recognition. In the United States one of the longest-standing such journals is the *Duke Mathematical Journal* (1935–   ). In the United Kingdom *Mathematika* (1954–   ), published by the Department of Mathematics of University College, London, springs immediately to mind. These are straightforwardly journals of research articles different in character from the periodical publications of seminar or lecture note material considered in Section 1.4.

2.7   JOURNALS FROM COMMERCIAL PUBLISHERS

There are two features of particular note which are more characteristic of commercially produced journals than of journals stemming from universities or learned societies. The first is the tendency to replace the refereeing system which precedes acceptance of a paper by a system in which a paper can be accepted by any one of a board of editors. The latter system may involve the paper in as rigorous examination as a full refereeing system; obviously much depends on the conscientiousness of the editor and the extent of his competence in the narrow field of the paper. But it should lead to quicker publication, if only because the hunt to find a competent referee is avoided. An example of this type of journal is *Inventiones Mathematicae* (1966–   ), published by Springer. It is the declared intention of this journal to publish articles within four months of acceptance, although the analysis of publication delays in the *American Mathematical Society. Notices,* **23**(5) (August 1976) show it to have fallen some way behind this. The refereeing system is, of course, not the only cause of delay in publication. Another is the sheer volume of material awaiting publication for a journal of standing which publishes regularly. *Inventiones Mathematicae* avoids this latter problem by publishing irregularly when it has an appropriate amount of material.

Publication delay can be further reduced by reproducing an author's typescript by a photographic process rather than having it typeset. An example of this type of journal, again from Springer, is *Manuscripta Mathematica* (1969–   ).

It might be supposed that the replacement of the refereeing system would lead to a lowering of the standard of published articles. However, the comment was made in the symposium on the publication of mathematical literature referred to in Section 1.1 that there was no significant difference as regards quality between refereed journals and those where articles were communicated via the editors.

The second feature of commercial periodical publication is the growth in the number of periodicals devoted to particular areas of mathematics*. One of the earlier ones was *Topology* (Pergamon, 1962–   ). Of course, not all journals of this type are commercial productions: one thinks of the *Notre Dame Journal of Formal Logic* (University of Notre Dame Press, Indiana, 1960–   ) and the Applied Probability Trust's *Journal of Applied Probability* (1964–   ), published in association with the LMS. Nevertheless it remains true that the growth of specialist journals of this kind in the 1960s and 1970s is a phenomenon made possible primarily by commercial publishers. For example, the *Journal of Differential Equations* (Academic Press, 1965–   ) replaces *Contributions to Differential Equations* (1963–1964), published by Interscience under the auspices of the Research Institute of Advanced Studies, Baltimore, and the University of Maryland.

The appearance of titles in new areas has continued into the 1970s—for example: *Journal of Number Theory* (Academic Press, 1969–   ); *Geometriae Dedicata* (Reidel, 1972–   ); and *Linear and Multilinear Algebra* (Gordon and Breach, 1973–   ). This suggests that the concern expressed about such journals in the symposium on the publication of mathematical literature (cf. Section 1.1) has not proved justified. It has proved possible for these journals to keep to fairly well-defined areas, and to attract articles of a quality comparable with that of those going to society publications.

* Details of journals specialising in the areas covered in later chapters of this book will be found in those chapters.

This emphasis on the role of commercial publication in two relatively recent developments in periodical publication should not lead one to overlook the fact that the longest-established mathematical journals precede the publication of mathematical societies by a considerable margin. The *Journal für die Reine und Angewandte Mathematik* was founded in 1826 and is still published by de Gruyter, who took over the original printers, in Berlin. Both *Annals of Mathematics* (Princeton University Press, 1884–   ) and *Mathematische Annalen* (Springer, 1868–   ) have distinguished histories.

2.8 TRANSLATIONS

The mathematician whose mother-tongue is English or American has a considerable advantage when it comes to mathematical journals. English-language journals almost wholly restrict themselves to publication in English, and even those such as the *Journal of Symbolic Logic* (cf. Section 6.1) which explicitly cater for articles in French or German contain only a small proportion of foreign-language material—and make no concessions to foreigners in the form of foreign-language contents lists or summaries. Indeed, the practice of giving summaries in languages other than the principal language(s) of publication seems to be unknown in the mathematical field.

In Western Europe periodicals which are of more than purely local interest are usually published in a mixture of English, French and German. The *Commentarii Mathematici Helvetici* has already been mentioned (cf. Section 2.5) as doing so. The *Journal für die Reine und Angewandte Mathematik* (cf. Section 2.7) has welcomed foreign contributions from the start (in its early days they were in French or Latin); it remains predominantly in German. In the Scandinavian *Acta Mathematica* (Almqvist och Wiksell for the Institut Mittag-Leffler, 1882–   ) English predominates. Italy forms an exception to the general pattern, publishing much material in Italian.

Many Eastern European journals likewise make use of French or English for much of their publication: for example, *Fundamenta Mathematicae* (cf. Section 6.1). It would be misleading to suggest that this means that all the major work from these countries is known in the West. One suspects that there is much valuable material in, for example, mathematical logic in Polish. In the Eastern bloc Russia forms an exception; apart from a small minority of articles in the languages of the constituent Republics, Russian is universal. Many Russian journals have cover-to-cover translations; but even so, only a small proportion of the material published in Russia is appearing in English.

Many countries outside Europe and North America have adopted English for their major publication. Examples are the *Israel Journal of Mathematics* (Weizmann Science Press, 1963–   ), the *Japanese Journal of Mathematics* (National Research Council of Japan, 1924–   ) and the *Journal* (1949–   ) of the Mathematical Society of Japan. In South America Spanish or Portuguese is used, depending on the country. The situation may change as countries such as Brazil welcome more foreign (particularly American) mathematicians.

The following is a summary of translations other than the AMS and LMS translations mentioned in Sections 2.3 and 2.4:

Akademiya Nauk SSSR. Doklady. The pure mathematics section is translated as *Soviet Mathematics–Doklady* (American Mathematical Society, 1960–   ).

Akademiya Nauk SSSR. Izvestiya–Seriya Matematicheskaya. From Volume 31, translated as *Mathematics of the USSR–Izvestiya* (American Mathematical Society, 1967–   ).

Akademiya Nauk SSSR: Matematicheskii Institut imeni V.A. Steklova. Trudy. From number 74, translated as *Steklov Institute of Mathematics. Proceedings* (American Mathematical Society, 1966–   ).

Algebra i Logika, Seminar. Sbornik Trudov. From Volume 7, translated as *Algebra and Logic* (Consultants Bureau, New York, 1968–   ).

Differentsial'nye Uravneniya (Minsk). Translated as *Differential Equations* (Faraday Press, New York, 1965–   ).

Funktsional'nyĭ Analiz i ego Prilozheinya. Translated as *Functional Analysis and its Applications* (Consultants Bureau, New York, 1967–   ).

Itogi Nauki–Seriya Matematika. From 1966, translated as *Progress in Mathematics* (Plenum Press, New York, 1968–1972).

Matematicheskie Zametki. Translated as *Mathematical Notes* (Consultants Bureau, New York, 1968–   ).

Matematicheskii Sbornik. From Volume 72, translated as *Mathematics of the USSR–Sbornik* (American Mathematical Society, 1967–   ).

Moscow, Universitet. Vestnik. Seriya 1. Matematika, Mekhanika. The Matematika section is translated, from Volume 24, as *Moscow University Mathematics Bulletin* (Faraday Press, New York, 1971–   ).

Moskovskoe Matematicheskoe Obschchestvo. Trudy. From Volume 12, translated as *Moscow Mathematical Society. Transactions* (American Mathematical Society, 1965–   ).

Sibirskii Matematicheskii Zhurnal. From Volume 7, translated as *Siberian Mathematical Journal* (Consultants Bureau, New York, 1967–   ).

Ukrainskii Matematicheskii Zhurnal. From Volume 19, translated as
Ukrainian Mathematical Journal (Consultants Bureau, New York,
1968–   ).

2.9   MONOGRAPH SERIES

For the purposes of the International Serials Data System, a serial is 'a
publication, in printed form or not, issued in successive parts usually
having numerical or chronological designations and intended to be
continued indefinitely. Serials include periodicals, newspapers, annuals
(reports, yearbooks, directories etc.), the journals, memoirs, proceedings, transactions etc. of societies, and monographic series'. The
definition excludes works produced in parts for a period predetermined
as finite such as multivolume treatises. The inclusion of monograph
series serves to emphasise that they can be more than the sum of the
monographs which make them up: the knowledge that a book belongs
to a particular series can tell one something about its level; conversely,
the knowledge of the existence of a major series can send one to a
library catalogue to browse through the entries in that series to identify
works in a particular area.

The monograph series of the AMS and LMS have already been
mentioned (Sections 2.3 and 2.4). Two long-established monograph
series from publishers are: 'Annals of Mathematics Studies', from
Princeton, and 'Ergebnisse der Mathematik und ihrer Grenzgebiete',
from Springer. By far the largest such series (and therefore the one to
which the remarks in the previous paragraph particularly apply) is
Springer's 'Lecture Notes in Mathematics'. This was founded in 1964 as
the result of discussions between a group of mathematicians and
Springer to provide a place for, and give a much wider circulation to,
the category of material including lecture notes and seminar material
from university departments. As with *Manuscripta Mathematica* (cf.
Section 2.7), material is reproduced by photographic means from
typescript, both to keep down the cost of each issue, and to try to
preserve something of the informal nature attaching to the type of
material included. One of the most useful publications of 1975 must be
the index to the first 500 volumes, which is available free from
Springer. It contains a numerical listing with author and subject
indexes, as well as prefatory material about the series by the publisher
and the editors, A. Dold and B. Eckman, and notes on the preparation
of material for inclusion.

The success of the 'Lecture Notes in Mathematics' series has led
Springer to start Lecture Note series in other areas such as Physics,
Computer Science, Biomathematics and Economics and Mathematical

Systems, so that the mathematicians' tendency simply to refer to the Springer Lecture Notes is liable in some circumstances to cause confusion.

2.10 BOURBAKI

The *Éléments de mathématique* produced by a group of French mathematicians under the pseudonym of Nicolas Bourbaki is a project which is completely *sui generis.* It aims to cover a large part of pure mathematics from a completely uniform axiomatic viewpoint, requiring of its readers only that elusive quality, mathematical maturity. As the recent publication history of the *Éléments de mathématique* is fairly complex it may help the reader to have the following summary of what has been published: the material in the third edition, which is the most recent one for the bulk of the treatise, was published in fascicules of one or more chapters from a particular subject area or 'Livre'. The chapters for any particular 'Livre' were not necessarily published in order, as Reference 117, Chapter 9 shows. Similarly, the book 'Algèbre commutative' is made up as follows:

Chapters I and II	Fasc. XXVII	(1961)
Chapters III and IV	Fasc. XXVIII	(1961)
Chapters V and VI	Fasc. XXX	(1964)
Chapter VII	Fasc. XXXI	(1965)

The 36 fascicules of this edition, published by Hermann over the period 1952-1971, cover the areas:

Livre I   Théorie des ensembles
Livre II   Algèbre
Livre III Topologie générale
Livre IV   Fonctions d'une variable réelle
Livre V   Espaces vectoriels topologiques
Livre VI   Intégration
           Groupes et algèbres de Lie
           Algèbre commutative

Translations of three of the books have appeared from Addison-Wesley: *Theory of sets* (1968); *General topology,* 2 vols (1966); and *Commutative algebra* (1972).

In the new edition which started to appear in 1970 the publication of separate paper-bound fascicules has been abandoned and the material is being issued in hard-bound volumes. The following sections are published or in progress:

    E   Théorie des ensembles (1970)
    A   Algèbre (Vol. 1, 1970)
    TG   Topologie générale (2 vols, 1971, 1974)
    FVR Fonctions d'une variable réelle
    EVT Espaces vectoriels topologiques
    INT Intégration
    AC   Algèbre commutative
    VAR Variétés différentielles et analytiques
    LIE   Groupes et algèbres de Lie
    TS   Théories spectrales

The preface to the published volumes states that the first six are self-contained or rely only on results in earlier volumes. After that there is cross-referencing between volumes as necessary.

3

Reference Material

S. Brookes

3.1 ENCYCLOPAEDIAS

Encyclopaedias, whether general or specialised, attempt to present information in a form that is concise and easily accessible. A good encyclopaedia should be able to be used as a dictionary for definitions of terms, as a handbook for basic data and as a bibliography for the important works in the field. It should provide information intelligible to those unfamiliar with the field, although the specialised encyclopaedias may require a knowledge of the general subject area, and so may be too technical for the beginner.

An encyclopaedia is never as up to date as the current literature; this is not necessarily a problem with mathematics, since a proof or a theory is not invalid just because it is months or years old. Most encyclopaedias try to combat this by updating themselves by yearbooks, continuous revision, new editions, or combinations of these.

The best-known general encyclopaedia is the *Encyclopaedia Britannica*[1]. The 15th edition, published in 1974, is in a new three-stage format. The updating policy of the 14th edition was yearbooks, and continuous revision leading to reprintings within the edition. There is no policy stated for the 15th edition, but it is implied that it will be the same as for the 14th.

The three stages are:

(1) *Propaedia.* This gives a brief overview of wide subject areas, and

contains the 'Outline of Knowledge', a classification of the contents of the encyclopaedia. With each heading in this classification there is a list of references to the articles in the *Macropaedia* on, or referring to, that subject. The classification of mathematics, in Section 10, could be used as an index to the encyclopaedia, or to give the relationships between the subjects within mathematics.

(2) *Micropaedia.* This is arranged in alphabetical order and gives short and factual information on specific subjects. It also serves as an index to the Macropaedia via the cross-reference system.

(3) *Macropaedia.* This is arranged in alphabetical order and gives fuller information on broader subjects. A subject that appears in the *Micropaedia* may only appear in the *Macropaedia* as an item in an article under a more general heading—for example, Lie groups is under Algebraic Topology Fundamentals in the *Macropaedia*. The *Macropaedia* articles are signed and have bibliographies (one, at least, containing 1972 material).

Both the *Micropaedia* and the *Macropaedia* contain biographies of famous mathematicians, and succinct histories of the mathematical disciplines. The level, and terminology, would in some areas be above a beginner, but someone with a knowledge of mathematics should be able to cope in an unfamiliar field.

A general encyclopaedia which would be more useful to a beginner is *Chambers's Encyclopaedia*[2], the mathematics in this is at approximately school level. It is in alphabetical order with signed articles, and is updated by yearbooks (these have a section devoted to 'Science', but do not contain mathematics). There are cross-references in the text, but to cover all aspects of an area the index must be used as well.

The *McGraw-Hill Encyclopaedia of Science and Technology*[3] and *Van Nostrand's Scientific Encyclopaedia*[4] (in 15 volumes and 1 volume, respectively) are two well-known scientific encyclopaedias. However, for mathematics, neither of them equals the *Encyclopaedia Britannica* in breadth of coverage or depth of coverage.

The *McGraw-Hill Encyclopaedia* is arranged alphabetically, by broad subject area—for example Lie groups occur under Group Theory. Orientation within a broad subject is given by this arrangement, but, as keywords are not distinguished by, for example, type-face, specific information is hard to pick out. Also, as the cross-reference system is adequate rather than generous, all the relationships between subjects are not given.

There are three indexes provided. The first is a contributor index, the second a subject index and the third a list of article titles under major subject divisions (all of mathematics and statistics is under Mathematics). The articles are generally in the form of a definition followed by an explanation. There are short bibliographies, but history and biographies are only given incidentally. A beginner should be able to cope with both the terminology and the level of treatment. It is kept up to date with yearbooks.

Van Nostrand's, 4th edition published in 1968, will not be updated until the 5th edition is published. It is a series of alphabetically arranged, unsigned, articles that are often merely expanded definitions. Words in an article which are headings to articles elsewhere in the text are denoted by a heavier type-face. However, if a word is not in heavy type it does not mean it is not a heading (e.g. transformation in the article on Lie groups). There are some synonym referrals and some text-to-text referrals, but there is no general index. History and biography are treated incidentally and consist of little more than dates, names and places.

There are encyclopaedias that deal with mathematics alone. The oldest of these is the *Encyklopädie der Mathematischen Wissenschaften*[5] (1898-1935 and 1939-1959). This is a series of high-level, scholarly articles on particular subjects. The 2nd edition seems to have ceased, but although sections, particularly those dealing with applications of mathematics, will be out of date, the 'academic' areas are still useful. The latest published encyclopaedia on mathematics is *The Universal Encyclopaedia of Mathematics*[6], in one volume, published in 1964. It is a translation and adaption of *Meyers Rechenduden,* which was published by the Bibliographisches Institut in Mannheim in 1960. The *Universal Encyclopaedia* is rather more a handbook than an encyclopaedia as it is an alphabetical list of definitions, sometimes with worked examples, of precise mathematical subjects (e.g. geometric mean is included but not geometry). It thus has no history, no biography and no bibliography, but it does contain useful tables and a subject-divided list of formulae. Its use is further limited as it has no cross-references.

The *Mémorial des Sciences Mathématiques*[7], which has been running since 1925, is perhaps not an encyclopaedia in the accepted understanding of the word. It is a series of monographs, of high standard, on different subjects, by different authors. For any subject it is as up to date as the date of the last monograph on that subject in the series. The recently planned *Encyclopaedia of Mathematics and its Applications,* edited by G.C. Rota and being published by Addison-Wesley, is again an open-ended series with each volume covering a particular field (Vol. 1, *Integral geometry and geometric probability,* by L.A. Santalo, and Vol. 2, *The theory of partitions,* by G.E. Andrews, have so far appeared).

The *Mathematisches Wörterbuch*[8] is another useful work that is difficult to class; it is more than a dictionary and not quite an encyclopaedia. It is an alphabetical arrangement of entries which range from simple definitions to thorough explanations. It contains no index, but has an extensive system of cross-references. It contains a list of symbols arranged by subject, brief biographies with bibliographies for better-known mathematics, and some history of mathematics.

3.2 HANDBOOKS

If you require concise information on a practical aspect of mathematics, a handbook may meet your need. A handbook should give facts, data, techniques, principles and definitions. Handbooks started as books (which could be carried) of essential information, but they have expanded and now some are multivolume. The four described below are all one-volume works, although Merritt has 378 pages and Korn and Korn has 1130. They are all arranged in broad subject chapters and are further subdivided within these, and all have subject indexes. This subject arrangement eliminates much of the necessity of stating relationships between subject areas.

Korn and Korn[9] is a good example of a mathematical handbook. The breakdown for each chapter is listed at the beginning of the chapter, and at the end there is a bibliography and a list of related topics treated in other chapters. The treatment is at quite a high level, with major terms and definitions, and more advanced information, denoted by different type-faces. There are, in addition to the subject index, cross-references within the text. It also contains lists of formulae and symbols, and tables.

Merritt[10] is a simpler handbook; it aims at giving the method, definition, etc., without the theory involved. Thus, for each subject there is a definition, an illustration where applicable and sometimes a worked example. If a method is used in more than one area, then it is described in each area. The subject index is augmented by cross-references in the text, and each chapter has a short bibliography at its end.

Meyler and Sutton[11] is of similar length and aims to give information from GCE to Honours degree level. It is divided into Pure Mathematics and Physics, and even the subject index is divided this way. It is most useful as a source of precise details on a specific subject, as it is virtually linked definitions and has neither cross-references nor bibliographies. It uses different type-faces to aid the quick recognition of definitions and laws. It contains useful lists of integrals, sums of series, etc.

Kuipers and Timman[12] is slightly different, as it is a collection of chapters written by different authors on specific subjects, beginning with a short history chapter. It, too, aims to provide 'fundamental mathematical knowledge' to people using mathematics and with some familiarity with mathematics. It is little more than definitions fleshed out into prose. There are no cross-references, and although the introduction states that there is a list of textbooks, the edition studied did not contain one.

3.3   ABSTRACTING JOURNALS

In an ideal world an abstract would appear quickly after the publication it abstracts. In the practical world this does not happen and, given the constraints of money and time, it may be impossible.

Mathematical Reviews[13] has been published since 1940 by the American Mathematical Society, with support by other bodies (e.g. National Science Foundation, London Mathematical Society). It is laid out in classification order. The classification is given in the last issue each year, and is capable of fine subdivisions, but *Mathematical Reviews* uses only the major divisions. This means that, if you are searching for material on a fine subject, you must search all the material listed under a broader subject to find it. It is indexed by author, in each issue and in each of the two volumes per year, and there are now cumulative author indexes for Vols 1-20 (1940-1960), Vols 21-28 (1961-1964) and Vols 29-44 (1965-1972). Each volume also has a 'key' index, which is an index by title of those works without authors or editors. Since Vol. 45 (1973), each volume has had a subject index: a listing by class mark giving the author and abstract number of the works classified there.

It covers, as well as periodicals (those covered are listed in the index issues), books, conferences, etc., and while these appear to be taken internationally, like the journals, no policy is stated for how, or from where, they are chosen.

The bibliographical details for each entry are in the original language, with Cyrillic, etc., Romanised, and sometimes translated. The abstracts themselves are mainly in English, French or German, and an English article may have an abstract in, for example, German, so a basic knowledge of these three languages is necessary to use *Mathematical Reviews* fully. The abstracts are signed, and, if taken from another source (e.g. *Computing Reviews*), this source is indicated. They are critical in the main, and sometimes quote from the preface or contents list.

Its major drawback is the time delay of approximately one year between the publication and its abstract appearing. In order to bridge this gap, the American Mathematical Society also publish *Current Mathematical Publications*[14], previously called *Contents of Con-*

temporary Mathematical Journals. This is a list of works which *Mathematical Reviews* has received in, usually, two weeks, which they intend to review. It is arranged as *Mathematical Reviews,* by major classification divisions, with author indexes, cross-references in the text for joint authors, key indexes, lists of journals covered and lists of new journals. The February 1975 issue contained mainly late 1974 material, so it is quite successful in its aim.

Another bridging publication of the American Mathematical Society is the *Index of Mathematical Papers*[15]. This states it covers 'articles processed during the preceding six months'. At the end of 1973 the policy for the *Index* changed, and it will now cover not only the journals but also the books, etc., which have been reviewed. As it will give the location of the reviews, it will also be able to be used as an index to *Mathematical Reviews.* It is in two parts: (1) Author index giving, alphabetically by author, the full bibliographical details of the works. (2) Subject index giving, in classified order, the authors and titles of the works. In this a P or an S denotes whether this classification is the primary or secondary one for the work. It is published each year in one volume of two parts; each part covering six months. The new coverage will probably delay it somewhat and therefore make it less useful as a bridging device, but the three publications, *Mathematical Reviews, Index* and *Current Mathematical Publications,* taken together have a comprehensive and up-to-date coverage of the mathematical literature.

The *Zentralblatt für Mathematik und ihre Grenzgebiete*[16] is also approximately one year behind. It is arranged by subject classification, and indexed by author, key and subject, i.e. class mark. Each tenth volume is a cumulative author index to the previous nine. It appears to have international coverage in periodicals, books, conferences, etc., and the reviews may be in English, French or German. The abstracts are signed, and their provenance is indicated; *Zentralblatt* takes these from *Mathematical Reviews,* etc., and also from the works themselves sometimes.

Bulletin Signalétique[17], published by the Centre National de la Recherche Scientifique since 1956, consists of many separate series, the first of which, no. 110, covers mathematics. Until 1971 this was called 'Mathématiques Appliquées, Informatique, Automatique'; it is now called 'Informatique, Automatique, Recherche Opérationelle, Gestion'. Although mathematics has been dropped from the title the subject classification headings have not changed; in fact, in 1973 the heading 'Thèses et Ouvrages Fondamentals aux Mathématiques' was added. There is one volume per year of twelve issues, the twelfth being the index issue. Each abstracting issue has an author index. There is an annual subject index, in two sections, which links key words with abstract numbers, and an annual author index which unfortunately does not give titles as the author index to *Mathematical Reviews* does. There is also an annual 'classification' index which gives the page at which each section of the classification schedule begins for each issue.

The abstracts are in French, and are indicative, but the bibliographical details are in the language of origin. Where the original language is in non-Roman script, the details are translated, and, for example, 'En russe' is given. It covers periodicals, and some conferences, theses, 'brevets' and reports, as received by Centre National de la Recherche Scientifique. It is, like the other abstracting journals, approximately one year behind.

Referativnyi Zhurnal. Matematika[18] is more than the major abstracting journal for material from the Russian bloc, as it aims at a comprehensive coverage of mathematics journals. A Russian reader can use it easily, possibly more easily than an English reader can use *Mathematical Reviews,* as a Russian title is given regardless of the original language, and all the abstracts are in Russian. However, to a non-Russian reader it seems forbidding, but a little patience, or the use of Copley[19], can make it a useful tool.

The bibliographical citations are in the original language, except oriental languages, which are changed into Cyrillic. For Roman-script items there is a Roman author index (in addition to the Cyrillic author index), which also lists joint authors, giving the abstract number and the subject classification. This is cumulated for each year. There is a classified subject index, with the headings in Russian, which is also cumulated each year. Although there is the same systematic arrangement, by subject, of the abstracts in each issue, this arrangement is not the same as the subject classification in the index.

A non-Russian reader can use it as a source of bibliographical details on works of a particular author via the Roman author index; or he can, if he knows a work on a subject, use the Roman author index to find its subject classification, and then the subject index to find the abstracts of other works classified there. Alternatively he can use an English–Russian dictionary to find the Russian equivalent of his subject, and then use the subject index.

The *Bibliografia Matematica Italiana*[20] began in 1951, and is not strictly an abstracting journal. It lists for each year those items published in mathematics in Italy, for that year. The 1970 volume was published in 1972; so it, too, has the long time lag. It gives just the bibliographical details, arranged alphabetically by author, and is accessed by a subject index, and by a list of broad subject headings (e.g. geometria, matematica applicata) with the authors listed beneath. It has ten-year cumulative indexes.

A useful bibliographical tool, although not strictly mathematical, is

Science Citation Index[21]. This is compiled by computer from the citations from approximately 2300 journals; it does not cover difficult-language journals very well. It can be used for tracing articles citing a particular work, and thus tracing works in similar fields, for tracing works by a particular author, for tracing works published 'from' an organisation by using the corporate index, and for a subject search using the Permuterm index. The three sections are:

(1) The index itself, arranged alphabetically by author's name with, for each publication, journal abbreviation, volume, year and pages, followed by the names of those who cited it.

(2) The source index, which contains alphabetically the names of the 'citers', and the brief titles and bibliographical details of their articles.

(3) The Permuterm index, which is an alphabetical list of keywords taken from the titles of the 'citers' articles. Each term is paired with other keywords from the title, so that if you are searching for a subject that has two keywords, you can easily find which titles contained both, and then, via the authors' names and the source index, the fuller title and the bibliographical details.

3.4 BIBLIOGRAPHIES

Mathematics is such a large and wide-ranging subject that bibliographies, in the usual meaning of a single work covering most of the literature of a subject, do not exist. There are, however, bibliographies that are restricted in time, e.g. Dick[22], or to a language, e.g. La Salle and Lefschetz[23], or to a particular aspect of mathematics, e.g. Gould[24].

One can, with a little license, consider catalogues and ceased abstracting journals as bibliographies, as they cover all of mathematics and all types of literature, although they are limited by time.

The most famous catalogues are those of the Royal Society of London. Between the *Catalogue*[25] and the *International Catalogue*[26] they cover 1800-1919. The *Catalogue* is arranged alphabetically by author and is the standard index to scientific papers from the transactions and proceedings of scientific societies, and scientific periodicals between 1800 and 1900. A seventeen volume subject index was planned for this, to be arranged by the subject classification used in the *International Catalogue*. Although only the first three, Mathematics, Mechanics and Physics, were finished, a mathematical subject search is possible for 1800-1900. The *International Catalogue,* while covering the same type of material, was issued annually in 17 subject 'volumes'. Each of these consists of a schedule of the classification, a list in classified order of the works covered, a similar list alphabetically by author and a subject index in English, French, German and Italian.

Jahrbuch über die Fortschritte der Mathematik[27] also provides access to nineteenth century mathematics. It was published from 1868 to 1942 as an abstracting journal. Its reviews are long, signed and in German. There are no indexes, but the arrangement is by subject classification, so some access is possible. As there was one volume per year and there are no cumulations, it would not be easy to use.

Series iii A/1, Mathematik, mechanik, astronomie, of the *Inhalts-verzeichnisse Sowjetischer Fachzeitschriften*[28], published from 1951 to 1965, abstracted Soviet mathematics, mechanics and astronomy. It is arranged in no easily discernible order, with no indexes, as it is in the form of joined-together cards with subject number and subject headings on the cards (so that they can be filed in subject order?). The abstracts, in German, are both signed and dated, and where necessary the bibliographical details have been both Romanized and translated into German.

3.5 BIOGRAPHY

If information is needed about a mathematician, the approach will depend on the information required. For living mathematicians, if a contact point for a particular person, or a list of who is where, is needed, the best source is the *World Directory*[29]. The latest edition is 1970, so it is not up to date, but it would give at least a start to a search. It is an alphabetical list, by name, which gives the surname, initials or christian names, titles or function (e.g. Dr or Lecturer), and an address (usually a 'business' address). It has a geographical index which is alphabetical by country, and lists the names of the mathematicians in that country.

One could wish that it was practical for the *World Directory* to contain the amount of information supplied by the *Directorio Latinamericano de Matematicas*[30]. This contains an alphabetical list of mathematicians giving names, sometimes dates, qualifications, present functions, addresses, and special fields of interest. Also, it has a geographical list of institutions giving addresses and qualifications issued; and alphabetical list of societies giving address and major personnel; an alphabetical list of mathematical publications giving details of frequency, availability, etc.; and a geographical index of people, institutions and publications.

If fuller information about a specific mathematician between 1858 and 1953 (at the moment) is needed, then Poggendorff[31] is a useful source. This is alphabetical by mathematician and contains short biographical notes followed by a list of his publications. Poggendorff not only covers mathematicians; it also attempts to include the important men in the other physical sciences as well.

If full biographical information is required and if the mathematician is sufficiently famous, then a biography or autobiography, or in some cases a biography from an encyclopaedia, is the best source.

3.6   DICTIONARIES

There are two types of dictionaries likely to be used by someone working in mathematics: monolingual ones to be used to elucidate a term and multilingual ones to be used for a foreign language. For monolingual dictionaries it is well to remember that the general ones (e.g. *Oxford, Websters'*) can be useful in this area. They may give different information from a specialised dictionary (e.g. the origins and early uses of words) but this does not make them less useful. It may be that professional mathematicians do not use such dictionaries, even to help in unfamiliar fields, as the level of the following dictionaries is no higher than undergraduate.

James and James[32] is quite up to date, considering its publication date of 1968, and is of undergraduate level. It is alphabetical, and the definitions are rather fuller than they strictly need to be. It does not give the origins of words, but does give their parts of speech and, where it is not obvious, their plurals. Synonyms are treated by referring to the defined term, and there are cross-references to related definitions. There are names and dates given for some of the famous mathematicians, but only where their name would occur as a mathematical term. It has an index in French, German, Russian and Spanish which gives the English term that is defined. It contains a list of symbols arranged by subject, and some commonly used tables.

Karush[33] aims at the high school and college student, but also contains definitions of terms from more advanced mathematics. It, too, is arranged alphabetically, but it consists of expanded definitions with important terms in heavier type. The cross-references are either explicit or denoted by different type-face. It gives brief histories of the major branches of mathematics (e.g. algebra, geometry). There is an appendix giving the names, dates, nationalities and brief descriptions of the 'Famous Mathematicians'. There is a list of references in another appendix, to which the definitions refer. While this is a most useful dictionary it is of a slightly lower level than James and James.

Another dictionary, which is of a lower level again, is Baker[34], which professes to be 'suitable for use up to degree standard, though equally valuable for less advanced study'. It is made up of an alphabetical arrangement of mainly short definitions, although there are the occasional longer ones (e.g. circle, differentiation), so that it is useful as a quick information source. It gives the names, dates and brief histories of major mathematicians. It has no subject framework, not even cross-references, and gives no bibliographies. It does give useful tables, formulae and simple symbols. It is useful as a word definition list, but has little else to recommend it.

Freiberger[35] is of a higher level than Baker, but is more restricted in coverage. This 'defines the terms and describes the methods in the applications of mathematics to thirty-one fields of physics and engineering'. It is an alphabetical arrangement of very full definitions linked by cross-references which are either explicit or denoted by type-face. It is, as it claims, a practical work, and gives no history of mathematics or mathematicians and no bibliographies. It also gives no useful tables, which is surprising, but it does have French, German, Russian and Spanish indexes.

The monolingual dictionary with the best reputation for accuracy, level and coverage is *Iwanami*[36]. To non-Japanese readers the fact that is is mainly in Japanese characters means that it is prohibitively difficult to use, even though it has a combined English, French and German index. It also has a Russian index and an index to personal names. Walford[37] states that it contains, as well as definitions of terms, biographies of mathematicians, a formulae list, tables and an annotated list of serials.

If a mathematician is attempting to read a work in a language that is not his own, he must turn to a bilingual or multilingual dictionary. The type of dictionary will depend on what knowledge the reader has of the foreign language. If he has none at all, a mathematical dictionary may be of little use on its own, as it will give him the equivalent of mathematical terms, but may not give the equivalents of common terms. In fact, he may need a brief description of grammatical rules as well, so that he can put his translated terms into the correct sequence. Many of the mathematical dictionaries recognise both these needs. If, however, he has sufficient knowledge to pick out the common words and the structure of the sentence, and only needs the technical terms translated, a mathematical dictionary will be all he needs.

In either case the dictionary used should be of the right time period, as terminology changes. Also, a good general dictionary (e.g. *Harrap's New Standard French and English Dictionary*[38]) may do if the subject being studied is not too specialised. An English mathematics dictionary which has indexes in the foreign language (e.g. James and James) may be sufficient.

As there are a great many dictionaries for translating a term from one language to another, the following list is a brief explanation of the major points of some of the mathematical ones.

3.6.1   German–English

MacIntyre and Witte[39] is an alphabetical list giving the English equivalent term of the German. It contains a brief grammar, two passages with their translations, a list of abbreviations and a list of German type-faces and script. It aims to help by containing the basic non-mathematical words as well. Its mathematics terms are restricted to pure mathematics.

Klaften[40] is both English to German and German to English. It is arranged in eight major subject groups (e.g. algebra, trigonometry) and then alphabetically, with a section at the beginning for fundamental terms. It gives model usages, and their translations within the text, and a separate abbreviations list. It would be much more useful if it had a term index covering all the sections.

Hyman[41] is an alphabetical list giving the English equivalent of the German terms with their parts of speech, and model usages and their translations. It has a list of abbreviations. It is restricted to mathematics terms.

Herland[42] is in two volumes, German to English and English to German. It is arranged alphabetically, and gives parts of speech, model usages and their translations, and various translations where necessary (e.g. balance, bank and equilibrium). It also gives cross-references, which is very useful. It is restricted to mathematics, but this includes, among others, logic and statistics. If a word has a common and a mathematical meaning, only the latter is given.

Meschkowski[43] is an alphabetical list giving the equivalent to a German term in English, French, Russian and Italian, with indexes from these languages to the German.

3.6.2   Russian–English

Milne-Thomson[44] is an alphabetical list giving the English equivalent of the Russian term with the part of speech, root of the Russian word and its meaning, and any other information necessary to the understanding of the term. It contains an outline of Russian grammar. It is functional, as it grew from the index Milne-Thomson developed to help himself translate mathematics, so it also contains non-mathematical terms.

Lohwater and Gould[45] is an alphabetical equivalent list, with parts of speech. It is preceded by a brief grammar. It was prepared under the auspices of the National Academy of the USA, the Academy of Sciences of the USSR and the American Mathematical Society, and aims at enabling the translation of Russian material in the area of mathematics and theoretical physics to be done using it alone. It has a Russian–English equivalent in Alexandrov[46].

Kramer[47] is an alphabetical equivalent list with a list of Russian abbreviations and symbols.

Burlak and Brooke[48] is an alphabetical equivalent list of mathematical and basic words. It also contains a short grammar and a specimen translation.

Recnik Matematickih Termina[49] is an alphabetical list giving the equivalent to a Russian term in Serbo-Croat, French, English and German, with French and German genders, and with indexes from these languages to the Russian.

3.6.3 French–English

Lyle[50] is two alphabetical lists, French to English and English to French. It gives the terms, pronunciations, parts of speech, equivalents, brief model usages and synonyms where applicable. It gives a definition where an equivalent term does not exist, and has cross-references. It contains a list of abbreviations, a list of numbers, a list of symbols by subject and sample translations. It states that it is up to undergraduate level, and seems to be a very useful dictionary.

There are also dictionaries for the more 'unusual' languages–for example, the following three.

De Frances[51] gives the equivalent of a Chinese character, or set of characters, in English. To use this at all, at least some knowledge of the structure of characters would be needed.

Wiskundewoordeboek[52] is two alphabetical lists, English to Afrikaans and Afrikaans to English, with parts of speech and synonyms. It covers only mathematics.

Gould and Obreanu[53] is an alphabetical list of the English equivalents to Roumanian terms. It also contains a simplified grammar.

REFERENCES

1. *The new encyclopaedia britannica,* 30 vols, 15th edn, William Benton for Encylopaedia Britannica Ltd, Chicago (1974)
2. *Chambers's encyclopaedia,* 15 vols, new rev. edn, Pergamon, Oxford (1973)
3. *McGraw-Hill encyclopaedia of science and technology,* 15 vols, McGraw-Hill, New York (1966)
4. *Van Nostrand's scientific encyclopaedia,* 4th edn, Van Nostrand, Princeton (1968)
5. *Encyklopädie der mathematischen Wissenschaften, mit Einschluss ihrer Anwendungen,* 1st edn (1898-1935), 2nd edn (1939-1959), Teubner, Leipzig
6. *The universal encyclopaedia of mathematics,* Allen and Unwin, London (1964)

7. *Mémorial des sciences mathématiques,* Gauthier-Villars, Paris (1925-   )
8. Naas, Josef and Schmid, Herman Ludwig, eds., *Mathematisches Wörterbuch; mit Einbeziehung der theoretischen Physik,* 2 vols, 3rd edn, Akademie Verlag, Berlin
9 Korn, Granino A. and Korn, Theresa M., *Mathematical handbook for scientists and engineers,* 2nd edn, McGraw-Hill, New York (1968)
10. Merritt, Frederick S., *Mathematics manual; methods and principles of the various branches of mathematics for reference, problem solving, and review,* McGraw-Hill, New York (1962)
11. Meyler, Dorothy S. and Sutton, O.G., *A compendium of mathematics and physics,* English Universities Press, London (1958)
12. Kuipers, L. and Timman, R., eds., *Handbook of mathematics,* Pergamon, Oxford (1969); English translation edited by I.N. Sneddon
13. *Mathematical Reviews,* American Mathematical Society, Providence, R.I. (1940-   )
14. *Current Mathematical Publications,* American Mathematical Society, Providence, R.I., 7- (1975-   ); continues *Contents of Contemporary Mathematical Journals,* 1-6 (1969-1974)
15. *Index of Mathematical Papers,* American Mathematical Society, Providence, R.I. (1970-   )
16. *Zentralblatt für Mathematik und ihre Grenzgebiete,* Springer, Berlin (1931-   )
17. *Bulletin Signalétique,* Centre National de la Recherche Scientifique, Paris (1956-   )
18. *Referativnyi Zhurnal. Matematika,* Viniti, Moscow (1953-   )
19. Copley, E.J., *A guide to Referativnyi Zhurnal,* 3rd edn, Science Reference Library, London (1975)
20. *Bibliografia Matematica Italiana,* Edizioni Cremonese, Rome (1951-   )
21. *Science Citation Index,* Institute for Scientific Information, Philadelphia (1961-   )
22. Dick, E.M., *Current information sources in mathematics; an annotated guide to books and periodicals 1960-1972,* Libraries Unlimited, Littleton, Colo. (1973)
23. La Salle, J.P. and Lefschetz, S., eds., *Recent Soviet contributions to mathematics,* Macmillan, New York (1962)
24. Gould, H.W., 'Research bibliography of two special number sequences', *Mathematica Monongaliae,* 12 (1971)
25. Royal Society of London, *Catalogue of scientific papers,* 1st series, 6 vols, 1800-1863, HMSO, London (1867-1872); 2nd series, 2 vols, 1864-1873; 3rd series, 3 vols (1874-1883) and supplement (1800-1883), Clay, London (1877-1902); 4th series, 7 vols (1884-1900), Cambridge University Press, Cambridge (1914-1925)
26. Royal Society of London, *International catalogue of scientific literature,* Harrison, London (1902-1921)
27. *Jahrbuch über die Fortschritte der Mathematik,* de Gruyter, Berlin (1868-1942)
28. *Inhaltsverzeichnisse Sowjetischer Fachzeitschriften. Serie iii A/1. Mathematik, Mechanik, Astronomie,* Institut für Dokumentation, Deutsche Akademie der Wissenschaften, Berlin (1951-1965)
29. *World directory of mathematicians,* 4th edn, Almqvist och Wiksell for the International Mathematical Union, Stockholm (1970)
30. *Directorio latinamericano de matematicas,* Unesco, Montevideo (1967)
31. Poggendorff, J.C., *Biographisch-literarisches Handwörterbuch zur Geschichte der exacten Wissenschaften,* J.A. Barth (varies), Leipzig (1858-   )

32. James, G. and James, R.C., *Mathematics dictionary*, 3rd edn, Van Nostrand, Princeton (1968)
33. Karush, William, *The crescent dictionary of mathematics*, Macmillan, New York (1962)
34. Baker, C.C.T., *Dictionary of mathematics*, Newnes, London (1961)
35. Freiberger, W.F., ed., *The international dictionary of applied mathematics*, Van Nostrand, Princeton (1960)
36. *Iwanami sugaku-jiten*, 2nd edn, Iwanami Shoten Publishers, Tokyo (1968)
37. Walford, A.J., ed., *Guide to reference material. Vol. 1, Science and technology*, 3rd edn, Library Association, London (1973)
38. *Harrap's new standard French and English dictionary. Part 1: French–English* 2 vols, rev. edn, Harrap, London (1972)
39. MacIntyre, Sheila and Witte, Edith, *German–English mathematical vocabulary*, 2nd edn, Oliver and Boyd, Edinburgh (1966)
40. Klaften, E. B., *Mathematical vocabulary*, Wila Verlag für Wirtschaftswerbung Wilhelm Lampl, Munich (1961)
41. Hyman, Charles, ed., *German–English mathematics dictionary*, Interlanguage Dictionaries, New York (1960)
42. Herland, Leo, *Dictionary of mathematical sciences*, 2 vols, 2nd edn, Harrap, London
43. Meschkowski, Herbert, *Mehrsprachenwörterbuch mathematischer Begriffe*, Bibliographisches Institut, Mannheim (1972)
44. Milne-Thomson, L.M., *Russian-English mathematical dictionary*, University of Wisconsin Press, Madison (1962)
45. Lohwater, A.J. and Gould, S.H., *Russian–English dictionary of the mathematical sciences*, American Mathematical Society, Providence, R.I. (1961)
46. Alexandrov, P.S. *et al.*, *Anglo-Russki slovar' mathematicheskikh terminov'*, Izdatel'stvo Inostrannoi Literatury, Moscow (1962)
47. Kramer, A., *Russian–English mathematical dictionary*, The author, Trenton, N.J. (1961)
48. Burlak, J. and Brooke, K., *Russian–English mathematical vocabulary*, Oliver and Boyd, Edinburgh (1963)
49. Matematicki Institut, Beogard, *Recnik matematickih termina*, Zavod za Izdavanje Udzbenika (1966)
50. Lyle, William David, *French and English dictionary of mathematical vocabulary*, Didier, Ottawa (1970)
51. De Frances, John, *Chinese–English glossary of the mathematical sciences*, American Mathematical Society, Providence, R.I. (1964)
52. Suid Afrikaanse Akademie vir Wetenskap en Kuns, *Wiskundewoordeboek. Mathematical dictionary*, Tafelberg, Johannesburg (1971)
53. Gould, S.H. and Obreanu, P.E., *Romanian–English dictionary and grammar for the mathematical sciences*, American Mathematical Society, Providence, R.I. (1967)

4

Mathematical Education

*A.G. Howson**

4.1 INTRODUCTION

For the purposes of this chapter, mathematical education may be defined as comprising attempts:

(1) to understand how mathematics is created, learned, communicated and taught most effectively;

(2) to design mathematical curricula which recognise the constraints induced by the students, their society, and its educational system;

(3) to effect changes in curricula (which we take to include content, method and procedures for evaluation and assessment);

(4) to foster within the population at large a growth of mathematical activity and an appreciation of the role of mathematics and of its nature.

It is a consequence of such a definition, of the great importance attached to mathematical education throughout the world today and of the consequent activity within that field, that any article of this nature must of necessity leave many gaps. In particular, to keep the bibliography within bounds, specific papers are referred to only if they are of a survey type or contain extensive or particularly valuable lists of references and suggestions for further reading. Again full particulars of

* The author would like to acknowledge his gratitude to those who commented so helpfully on an earlier draft of this chapter.

the complete works of certain major authors, projects, associations, etc., have not been included. In such cases one has attempted merely to draw the reader's attention to the existence of a body of important material.

4.2   JOURNALS AND INDEXING PERIODICALS

It is erroneous to think of mathematical education as a new study or activity. There is a long history of attempts to establish mathematics as a constituent part of any general education, and to convince society at large of the part which mathematics plays and could play in its life. For well over a century periodicals in Britain have carried articles on mathematical education and the *Mathematical Gazette,* a journal devoted to the improvement of mathematics teaching, is now over 80 years old. It is still through periodicals that one can learn most quickly of the different developments and research activities taking place within mathematical education. The number of journals which have been established within the last few years—some on an international rather than national basis—bears witness to the great upsurge of interest and activity in this field.

Of the many journals which publish papers and articles on mathematical education some are entirely devoted to mathematics and its associated educational problems, others have wider objectives and contain only the occasional (but nevertheless valuable) article on mathematical education.

Of the former, the reader's attention is particularly drawn to the following (the year of foundation being given in parentheses):

1. *Educational Studies in Mathematics* (Reidel, Netherlands; 1968)
2. *International Journal of Mathematical Education in Science and Technology* (John Wiley, UK/USA; 1970)
3. *The Mathematical Gazette* (Bell, for the Mathematical Association (MA); 1894)
4. *Mathematics in School* (Longman, for the Mathematical Association; 1971)
5. *Mathematics Teaching* (Association of Teachers of Mathematics (ATM); 1955)
6. *The Bulletin of the Institute of Mathematics and its Applications* (IMA; 1965)
7. *The American Mathematical Monthly* (Mathematical Association of America (MAA); 1894)
8. *The Mathematics Teacher* (National Council of Teachers of Mathematics (NCTM); 1908)

9. *The Arithmetic Teacher* (NCTM; 1954)
10. *Journal for Research in Mathematics Education* (NCTM; 1970)

Typical of the many journals published in languages other than English are:

11. *Didaktik der Mathematik* (Federal Germany)
12. *Bulletin de l'APM* (France)
13. *Periodico di Mathematiche* (Italy)
14. *NICO* (Centre Belge de Pédagogie de la Mathématique)

(These periodicals do not all attempt to cover every sector of mathematical education. Thus, for example, References 2, 3, 6 and 7 are more interested in problems relating to the mathematical education of those aged 16 or over.)

Other journals in which valuable articles on mathematical education are to be found include:

15. *Review of Educational Research* (American Education Research Association; 1931)
16. *British Journal of Educational Psychology* (Scottish Academic Press; 1931)
17. *Journal of Curriculum Studies* (Collins; 1968)
18. *Child Development* (University of Chicago Press; 1930)

and, of course, a host of others. (A list of journals devoted to mathematics and/or education is to be found in

19. *Ulrich's International Periodicals Directory*; cf. Section 2.1.)

In addition to these journals, the educational press frequently carries up-to-date accounts of work in action. For example, each year the *Times Educational Supplement* produces two special supplements devoted to mathematical education.

As an aid to the researcher, who cannot be expected to read solidly through the contents lists of all such periodicals, there are various indexing journals (although none, so far, that plays a corresponding role to the mathematical researcher's *Mathematical Reviews*).

For example, the titles and publication details of articles printed in a wide variety of British journals are to be found in

20. *The British Education Index* (British National Bibliography, London)

and in

21. *SATIS* (Science and Technology Information Sources, National Centre for School Technology, Trent Polytechnic, Notts.)

A list of references specifically related to mathematical education taken from 12 British journals and covering the period 1961-1974 has been published by the Centre for Studies in Science Education, Leeds University (22. D.C. Carter and G.T. Wain, *References of use to teachers of mathematics*). A cumulative index to *The Arithmetic Teacher,* covering the years 1954-1973, has been published by NCTM.
Indexing journals outside Britain include

23. *Zentralblatt für Didaktik der Mathematik* (Klett, GFR)
24. *Current Index to Journals in Education* (Macmillan, N.Y.) (see also 105)

Other sources of value are those which list theses for higher degrees, for example

25. 'Doctoral Dissertation Research in Science and Mathematics', *School Science and Maths* (annual abstracts)
26. *Index to Theses for Higher Degrees* (Great Britain and Ireland); Aslib
27. *Research in Science and Mathematics Education* (theses for higher degrees in British Universities, 1968-1971 and 1971-1973; Centre for Studies in Science Education, Leeds University)
27a. *Register of Education Research in the UK* 1973-1976 (National Foundation for Education Research, 1976-   )

4.3   ASSOCIATIONS AND INTERNATIONAL ORGANISATIONS

In addition to their journals, the various subject associations also produce a wide variety of reports, yearbooks, etc.
Of outstanding interest are the annual yearbooks of the National Council of Teachers of Mathematics. Recent titles include

28. *Historical topics for the mathematics classroom* (31st Yearbook; 1969)
29. *Instructional aids in mathematics* (34th Yearbook; 1973)
30. *The slow learner in mathematics* (35th Yearbook; 1972)
31. *Geometry in the mathematics curriculum* (36th Yearbook; 1973)
32. *Mathematics learning in early childhood* (37th Yearbook; 1975)

A yearbook of another US body—the National Society for the Study of Education—of particular interest is

33. *Mathematics education* (69th Yearbook; University of Chicago Press, 1970)

In Britain the MA and ATM have both published a series of books, reports and working papers, details of which can be obtained from the associations concerned.

Of special interest for its extensive bibliography is the ATM publication

34. T.J. Fletcher (ed.), *Some lessons in mathematics* (Cambridge University Press, 1964)

Since 1960 there have also been many important reports published by the international organisations OEEC (later OECD) and Unesco. Of these,

35. *New thinking in school mathematics* (OEEC, 1961)

and

36. *Synopses for modern secondary school mathematics* (OEEC, 1961)

merit attention because of the seminal part they played in the changes which followed.

The various ways in which these reforms were carried out in different countries are illustrated in

37. *New trends in mathematics teaching* (Unesco, Vol. 1, 1966; Vol. 2, 1968)

Volume 3 of *New trends* (1972) differs from its predecessors in that it does not reprint papers by individuals but presents a survey of developments in mathematical education based on the work of a group of educators gathered together from several countries. The work contains some particularly useful bibliographies on 'Geometry', 'Logic', 'Applications of Mathematics' and 'Research in Mathematical Education'.

The deliberations of a considerably larger meeting of educators are summarised in

38. *Developments in mathematical education* (Proceedings of the 2nd ICME; Cambridge University Press, 1973)

This report gives an indication of the activities of the various working groups set up at the Exeter Congress and also contains certain papers (notably by Freudenthal, Hawkins, Leach, Philp, Thom and Fischbein) which show the wide variety of ways in which study, research and developmental activity can take place within mathematical education. (The papers of one of the Congress's working groups were published separately:

39. E. Choat (ed.), *Preschool and primary mathematics,* Ward Lock Educational, 1973)

The papers given by invited speakers at the 1st ICME (Lyons, 1969) are to be found in *Educational Studies in Mathematics,* 2 (1969).

It is now planned to hold an International Congress on Mathematical Education every four years (in leap years) and they present the mathematical educator with a valuable opportunity to obtain information on what is happening in other countries. In other 'even-numbered years' the International Congress of Mathematicians is held and this always contains a section devoted to mathematical education. ICMI (the International Commission on Mathematical Instruction) has also helped promote international symposia on a variety of topics such as 'Applicable mathematics in secondary schools' and 'Mathematics and language in emergent countries'. The proceedings of these two symposia have been published:

40. *New aspects of mathematical applications at school level* (L'Institut Grand Ducal, Luxembourg, 1975)
41. *Interactions between linguistics and mathematical education* (Unesco, 1975)

4.4 THE DEVELOPMENT OF MATHEMATICAL EDUCATION

The study of how mathematical education has developed and of those factors that have affected it in the past has recently attracted considerable attention. This has resulted in the publication of works such as

42. *A history of mathematical education in the USA and Canada* (NCTM, 1970)
43. *The Mathematical Association of America: its first fifty years* (MAA, 1972)
44. E.G.R. Taylor, (a) *The mathematical practitioners of Tudor and Stuart England* (Cambridge University Press, 1954), (b) *The mathematical practitioners of Hanoverian England* (Cambridge University Press, 1966)

The worker in this area has, of course, a wide range of sources. A biobibliography of mathematical books published in Britain before 1850 is in the course of preparation by P.J. Wallis of the University of Newcastle-upon-Tyne (who, in the meantime, can provide various checklists of publications, e.g. of British 'Euclids'). Typical secondary sources of particular value for their bibliographies include

45. F. Watson, *The beginnings of the teaching of modern subjects in England* (1909; reprinted 1971 by S.R. Publishers)

46. A. de Morgan, *Arithmetical books* (1847; reprinted as an appendix to D.E. Smith's *Rara arithmetica,* Chelsea, 1970)
47. R.C. Archibald, 'Notes on some minor English mathematical serials' (*Mathematical Gazette,* xiv, 379-400, 1929)

Those interested in the development of a philosophy of mathematical education and of teaching methods will find starting points in such books as

48. A. de Morgan, Various essays extracted from the *Quarterly Journal of Education, The Schoolmaster* (1836)
49. B. Branford, *A study of mathematical education* (Oxford University Press, 1908)
50. C. Godfrey and A.W. Siddons, *The teaching of elementary mathematics* (Cambridge University Press, 1931) [Godfrey's chapters, written about 1910, are the ones of interest.]
51. G.St.L. Carson, *Essays on mathematical education* (Ginn, 1913)

Primary sources of particular value are

52. *Minutes of the Committee of Council in Education* (annual parliamentary papers describing schools and training colleges in the early years of Victoria's reign)
53. *Special reports on the teaching of mathematics in the United Kingdom* (2 vols; HMSO, 1912)

and that first attempt to persuade readers of English of the power and utility of mathematics:

54. J. Dee, *The mathematicall praeface* 1570; reprinted 1975 by Science History Publications, New York)

Reference 53 consists of papers prepared for the 1912 ICM held at Cambridge. The International Commission on Mathematical Education, which was established in 1908, collected together for the 1912 meeting an enormous number of reports from all over the globe. The result is an unparalleled survey of mathematical education throughout the world at one particular instant. The collected reports can be found in, for example, the library of the Mathematical Association at Leicester University.

A more compact survey written about that period and one which contains an impressive and useful bibliography is

55. J.W.A. Young, *The teaching of mathematics* (Longman, 1907; revised edition 1914)

4.5 SURVEYS

In recent years there have been several attempts to survey the whole field of mathematical education—sometimes by means of symposia.
The reader's attention is directed to

56. L.R. Chapman (ed.), *The process of learning mathematics* (Pergamon, 1972)
57. H.B. Griffiths and A.G. Howson, *Mathematics: society and curricula* (Cambridge University Press, 1974)
58. G. Wain (ed.), *Mathematical education* (Van Nostrand Reinhold; to appear)
59. H. Freudenthal, *Mathematics as an educational task* (Reidel, 1973)
60. W. Servais and T. Varga (eds), *Teaching school mathematics* (Penguin, 1971)

(References 57, 58 and 60 contain good bibliographies; Reference 56 is patchy but strong on psychology; Reference 59 is more a personal testament than a survey—it lacks an index and an adequate bibliography—yet it is a most stimulating book which contains many implicit suggestions for research topics.)

4.6 THE PSYCHOLOGY OF MATHEMATICS LEARNING

Readers of the books mentioned in Section 4.4 will note a growing and maturing interest in the manner in which children learn mathematics. In the early years of Victoria's reign Thomas Tate was pleading the case of 'faculty psychology', a theory that by Edwardian times (see Godfrey in Reference 50) had been ousted in popular favour by the Herbartian view of learning. (For a survey of this early work in educational psychology see also the articles by Burt and Hamley in 61. *Secondary Education* (Spens Report); HMSO, 1938.)

By the 1930s other learning theories were emerging, including the 'Gestalt theory'. In recent decades there have been many researches carried out within mathematical education in the field of educational psychology. It is impossible here to do more than list some of the key names: Ausubel, Bruner, Dienes, Gagné, Gattegno, Piaget, Skemp and Skinner—any university or equivalent library should contain a selection of these authors' works.

Surveys of work in psychology with a bearing on mathematical education are given by Peel in Reference 60 and by Shulman in Reference 33. Some useful articles and bibliographies are contained in References 56 and 58.

233987

Many surveys and critiques of Piaget's work are available, and we mention

62. E.V. Sullivan, *Piaget and the school curriculum: a critical appraisal*, OISE, 1967

since it draws attention to the Ontario Institute for Studies in Education, a body which from time to time publishes booklets of great interest to mathematical educators. Papers on psychology (containing good bibliographies) emanating from its Psychology of Mathematical Education Workshop are published by the Centre for Science Education, Chelsea College. Translations of Russian papers on psychology have been published by SMSG/A.C. Vroman under the title *Soviet Studies in the Psychology of Learning and Teaching Mathematics* (7 vols).

The problems of how children learn mathematics and of how mathematicians solve problems are interrelated, and it is therefore relevant here to draw attention to what is sometimes called the study of heuristics—the art of problem solving. The key book is still

63. G. Polya, *Mathematics and plausible reasoning* (2 vols; Oxford, 1954)
(64. G. Polya, *How to solve it,* Doubleday Anchor Books, 1957, though a more modest book, provides a good introduction to the author's arguments.)

A bibliography describing research in this field is contained in

65. M. Jerman, *Instruction in problem solving and an analysis of structural variables that contribute to problem solving difficulty* (Technical Report No. 180, 1971, Psychology and Education Series, Stanford University)

4.7 CURRICULUM DESIGN

The 'theory' of curriculum design is a comparative newcomer to the educational scene, yet guidelines for the construction of the curriculum have, of course, been laid down by mathematical educators for many generations.

A useful survey of curriculum design in a general educational setting (including good bibliographies) is

66. R. Hooper (ed.), *The curriculum* (Open University Press, 1971)

In the area of mathematical education one would wish to draw attention to References 56, 57 and 60, to the publications of the various mathematical associations (Section 4.3) and more especially to the work of the many projects that have been established during the last two decades.

Brief details of projects from many different countries can be found in the series of 'clearing house' reports issued by the University of Maryland. A description of projects on the continent of Europe is contained in

67. W. Servais, 'Continental traditions and reforms' (*International Journal of Mathematical Education in Science and Technology,* 6(1), 37-58, 1975)

The Mathematical Association has published reports describing mathematical projects in Britain. A list of some of the major projects is also to be found in Reference 57, together with the names of their publishers.

Curriculum development work is continuing in all areas of mathematical education. Thus in Britain, for example, the School Mathematics Project is currently working on a 7–13 Project, while at two extremes of the educational scale the Schools Council is sponsoring a project on the mathematical experiences of the pre-school child (Centre for Science Education, Chelsea College) and the London Mathematical Society has recently established a newsletter to keep its members informed of initiatives at university level.

The various investigations mounted by the Schools Council—the public body responsible for the curriculum and examinations in England and Wales—are described in 68. *Schools Council Project Profiles and Index,* which is revised at frequent intervals.

One exercise in curriculum design—an entirely theoretical one—merits especial attention:

69. CCSM, *Goals for school mathematics* (Houghton Mifflin, 1963)

Designing a curriculum is one thing; having it accepted in the classroom and taught in a meaningful ungarbled way is another. The manner in which projects can best work and the effectiveness of various means of dissemination are issues that would repay study in depth.

Two histories of major projects are available:

70. B. Thwaites, *SMP: the first ten years* (Cambridge University Press, 1972)
71. *SMSG: the making of a curriculum* (SMSG, 1965)

but these are somewhat cold, formal statements. More personal accounts of curriculum development work (albeit on a limited scale) are to be found in

72. *The Fife mathematics project* (Oxford University Press, 1975)

There is an extensive bibliography on the dissemination of ideas in Reference 58.

The problem of evaluating the work of projects is a difficult one. Useful books and papers on the topic include

73. American Educational Research Association (AERA), *Monograph series on curriculum evaluation* (Various titles; Rand McNally, 1967–   )
74. M. Parlett and D. Hamilton, *Evaluation as illumination* (Occasional Paper 9, Centre for Educational Sciences, University of Edinburgh, 1973)

(See also References 33, 72 and 81.)

Accounts of the evaluative work carried out by some of the Schools Council projects are contained in

75. *Evaluation in curriculum development* (Schools Council Research Studies, Macmillan, 1973)

The problems of evaluation and dissemination are, however, closely connected with the subjects to be mentioned in the following two sections.

4.8   ASSESSMENT AND EXAMINATIONS

A key work on assessment is

76. **B.S.** Bloom *et al., Handbook on formative and summative evaluation of student learning* (McGraw-Hill, 1971)

which contains a long section on the evaluation of learning in secondary school mathematics by J.W. Wilson, a member of the SMSG team concerned with the National Longitudinal Study in Mathematical Ability—an enormous evaluative exercise carried out in the US (see also Reference 33).

Wilson's article contains many references to work in the US and Canada. Further references can be found in the chapter on evaluation in Reference 37, Vol. 3.

Practical examples of attempts to examine mathematics in new ways can also be found in the examinations bulletins published from time to time by the Schools Council. The National Foundation for Educational Research (UK) has shown particular interest in the problems of evaluation and assessment and frequently publishes monographs of interest to the researcher.

Although in the past most effort has gone into the measuring of cognitive achievements in mathematics, there is now increasing interest in the problem of investigating students' *attitudes* to mathematics— factors in the affective domain. As a study of Wilson's paper will show, the measures so far devised leave much to be desired. Nevertheless, this is an important part of mathematical education: study of it has already given rise to research and it is likely to be the subject of many more investigations.

4.9   TEACHER TRAINING

The implementation of change depends on the flexibility and competence of the classroom teacher. Producing teachers who can successfully cope with change is a primary objective of all pre-service and in-service training work. In the US the MAA has prepared several reports on the training of teachers; in Britain the Association of Teachers in Colleges and Departments of Education (ATCDE) has produced booklets outlining the construction of programmes for intending teachers and suggesting possible solutions; the OECD and Unesco have, together with other bodies too numerous to detail, also produced reports.

The curriculum development projects of the 1960s gave rise to several alternative methods of dealing with in-service training: the short course, the teachers' centre, etc. There is a growing literature on this topic. Most of it is to be found in periodicals, in research papers and reports, and in theses, but the surveys mentioned in Section 4.5 do list some relevant books. The position of teacher training and of the institutions given over to it is, however, far from stable; neither is it obvious what financial resources are to be made available for it in the near future. Clearly, improvisation will once more be the order of the day, and those with research interests in this particular field will be particularly dependent on up-to-date reports and studies. A statement of 'what is and what might be' as far as the English scene is concerned is to be found in

77. *Teacher education and training* (James Report; HMSO, 1972)

Reports and articles concerning teacher training in Britain can be found in the journal established in 1974:

78. *Mathematical   Education   for   Teachers* (Homerton   College, Cambridge)

4.10 EDUCATIONAL TECHNOLOGY AND INDEPENDENT LEARNING METHODS

The increased availability of electromechanical apparatus (programmed learning machines, tape cassettes, slide projectors, film strips, and, above all, the computer and television), the views of certain educational psychologists on learning theory, attempts to bring the 'systems approach' of engineering to the classroom, and the coming of comprehensive schools with their emphasis on 'mixed-ability' teaching have combined, in varying ways, to make 'educational technology' a central issue in mathematical education today.

Programmed learning itself had a comparatively short stay in the 'top ten' of educational fashions. Its rise and (partial) fall is charted in

79. G.D.M. Leith, *Second thoughts on programmed learning* (NCET; Councils and Education, 1969)

This booklet is one of a series published by the Council for Educational Technology, CET (and its forerunner NCET). Other booklets in this series of interest—and possessing useful bibliographies— are on the computer in education, on proposals for a mathematics project that would make use of various techniques within educational technology (Howson and Eraut, 1969, which gave rise to the Continuing Mathematics Project at Sussex University) and on 'In- Service Education for Innovation' (Eraut, 1972). The Council is also responsible for the administration of several projects concerned with the use of the computer as an educational aid. This is, of course, a development of the pioneer work that was carried out in the US in the 1960s. Typical project descriptions from that era are

80(a). P. Suppes, 'The uses of computers in education' (*Scientific American,* 206-220, September, 1966), (b). P. Suppes *et al., Stanford's 1965-6 arithmetic program* (Academic Press, 1968)

Educational technology's greatest triumph in England, however, is to be found in the work of the Open University. Here various media have been exploited in the design of what is basically a 'super corres- pondence course'. An enormous quantity of materials has been prepared by the Open University. Research into the effectiveness of the methods employed and the success of the university's mathematical department in attaining its objectives—and, indeed, a critique of the soundness of those objectives—should prove of great interest and value.

In schools, educational technology's aid has been sought primarily to deal with the problems of mixed-ability teaching. The classic project in this area is that mounted by the Swedes—IMU (this was also the subject of a particularly interesting exercise in evaluation: 81. I. Larsson,

Individualised mathematics teaching, CWK Gleerup/Lund, 1973). This Swedish series is now available in an English adaptation (Caffrey Smith). Other courses based on individualised learning which make use of work cards and alternative materials have been produced by a number of publishers.

A useful bibliography on educational technology covering all aspects of education is

82. S. Stagg and M. Eraut, *A select bibliography of educational technology* (2nd edn, Council for Educational Technology, 1975)

The journal

83. *Programmed Learning and Educational Technology* (Sweet and Maxwell)

contains articles relating to mathematics teaching, as do the yearbooks produced by the Association for Programmed Learning and Educational Technology (APLET) and published by Kogan Page.

4.11 APPARATUS AND FILMS

The position with respect to the availability of apparatus, films, etc., is constantly changing. Film lists have been produced by the ATM from time to time, the most recent in 1975.

An older, but still useful, publication is

84. P.R. Burgraeve, *Report on films for the teaching of mathematics in Europe* (Strasbourg Council of Europe, Council for Cultural Co-operation, 1970)

New films and apparatus are reviewed in various journals (of the British mathematical journals, References 4 and 5 are best for this purpose), and a list of various sources of mathematical films can be found in Reference 38. Some of the problems faced by the designer of mathematical films are discussed in

85. *Films and film making in mathematics* (ATM, 1967)

An important new type of film-making makes use of the computer. Various teams—for example, at the Open University and at the London University Computer Unit—are attempting to exploit these new possibilities.

The development of small electronic calculators has been proceeding at a remarkable rate. As a result, any account of what models are available rapidly becomes out of date. Several of the journals mentioned in Section 4.2, however, provide frequent up-to-date surveys. Such calculators are almost certain to have a very profound effect on the teaching of mathematics in all sectors of education and their use on a wide scale will necessarily initiate a considerable amount of research and development work.

The question of how apparatus can be most effectively used in the classroom is one which belongs to the domain of educational psychology. The problem has been investigated by, among others, several of the authors mentioned in Section 4.6.

4.12 MATHEMATICAL COMPETITIONS AND OTHER EXTRACURRICULAR ACTIVITIES

For many years now Hungarian children have responded to the opportunity to pit their mathematical wits against those of others in annual olympiads. Gradually, the ideal of holding problem-solving contests has spread and there is today an International Mathematical Olympiad to which a dozen or more countries send teams. The story of the growth of these competitions together with an extensive bibliography can be found in

86. 'ICMI report on mathematical contests in secondary education' (*Educational Studies in Mathematics,* **2**, 80-114, 1969)

Some research has been carried out to see how successful these competitions are in picking out future mathematicians and in creating interest in mathematics. Little appears, however, to have been done on this in the UK, or, indeed, on investigating whether the similar objectives of the 'Oxbridge' scholarship examinations are attained.

In Russia great use has been made of Olympiads as detectors of latent mathematical ability. See for example,

87. S.L. Sobolev, 'Mathematical Olympiads in the Soviet Union' (*Proceedings of the Japan/ICMI Seminar,* 16-23; JSME, Tokyo, 1975).

The Russians are also thought of in connection with two other initiatives in mathematical education. One is the special school for the mathematically gifted; see, for example,

88. A. Owen and F.R. Watson, 'The mathematical boarding schools of the USSR' (*Mathematical Gazette,* **lviii**, 188-195, 1974)

The other is the establishment of mathematical clubs for students. An impressive list of papers on mathematical clubs of all kinds can be found in

89. W.L. Schaaf, *A bibliography of recreational mathematics* (3 vols, NCTM, 1970-1973)

The production of extracurricular and recreational material for students (of all ages) presents mathematical educators with a pressing problem. It is an old problem (as Reference 47 shows) which is met today in a variety of ways, most notably by Martin Gardner in the *Scientific American.* This last is but one way in which an attempt is made to foster mathematical activities among 'non-mathematicians'. The general problem of telling the public at large about what mathematicians do and why they do it has still hardly been touched on.

4.13   THE NEEDS OF SOCIETY AND, IN PARTICULAR, OF OTHER SUBJECTS

The curriculum designer, at whatever level he operates at, must be aware of the way in which students are likely to be called upon to use mathematics both in their other studies and in their future work.

Investigations have recently been carried out into some of these 'interface' problems and one would mention the following as possible starting-off points for future research:

90. R.E. Gaskell and M.S. Klamkin, 'The industrial mathematician views his profession' (*American Mathematical Monthly,* **81,** 699-716, 1974) [contains an excellent bibliography]
91. *Mathematical needs of school leavers entering employment* (IMA, 1975)
92. 'The mathematical needs of "A-level" physics students' (Royal Society/Institute of Physics; *Physics Education,* June 1973)
93. 'Report of the working party on mathematics for biologists' (Royal Society/Institute of Biology; *International Journal of Mathematical Education in Science and Technology,* **6,** 123-135, 1975)
94. 'Mathematics and school chemistry' (Royal Society/Institute of Chemistry; *Education and Science,* January 1974)

The production of examples that can be employed in day-to-day classroom teaching showing how mathematics is used is a valuable contribution to mathematical education. The Mathematical Association has produced reports giving applications of school mathematics. At a higher level the MAA sponsored the production of

95. B. Noble, *Applications of undergraduate mathematics in engineering* (Macmillan, New York, 1967)

The **IMA** has devoted several issues of its *Bulletin* (Reference 6) to descriptions of how mathematics is applied within commerce, industry and the sciences; for example,

'Mathematics in clinical medicine' (**10**, 1/2, Jan./Feb., 1974)
'Medical and biological applications of statistics' (**11**, 3/4, March/April, 1975)
'The mathematics of telecommunications traffic' (**11**, 5, May 1975)

Two bibliographies which illustrate how mathematics is used outside the physical sciences and engineering are

96. D. Thompson, 'A selective bibliography of quantitative methods in geography' (*Geography,* **54**(1), 74-83, Jan. 1969)
97. G.L. Baldwin *et al.,* 'A program in mathematical analysis for the life and management sciences' (*American Mathematical Monthly,* **825**, 514-520, May 1975)

4.14 COMPUTERS

The use of the computer as an educational aid has already been mentioned in Section 4.10. The computer, however, has a more crucial role to play in mathematical education—for its power is such that it has been able to influence mathematical thought and the nature of mathematical activity. Once again one can only report that the rate of development is such that no list of set references will remain of value for long.

However, as some indication of the varying ways in which the problems of mathematical education and computer education overlap, one might mention the work of Iveson to design a new *mathematical* notation more suited for a computer age, that of Papert (see Reference 38, for references), who used the computer to 'teach children to be mathematicians rather than to teach them mathematics', and the work of the SMP group on computing, who have suggested possibilities of using the computer to reinforce the learning of mathematics (handbooks published by the Cambridge University Press).

Various reports on computer education have been published from time to time. Typical examples are

98. *Computer science in secondary education* (OECD, 1971)
99. *World conference on computer education, Amsterdam, 1970* (IFIP Amsterdam, 1970)

and the double number **5** (3/4) (1974) of Reference 2 devoted to papers given at a conference on 'Computers in higher education'.

IFIP—the International Federation for Information Processing—held a second world conference at Marseilles in 1975 at which information on recent work within computer education was exchanged. It has also produced a number of working party reports and sets of guidelines for teachers. The report of the 1975 conference was published as

99a. *Computers in education* (North-Holland, 1975)

In Britain the British Computer Society has published a series of working papers on computer education. It also has issued a film list and book list, the latter being regularly updated. The Computer Education Group publishes a periodical, *Computer Education,* and the National Computing Centre has provided an advisory service for schools. In Scotland the Scottish Education Department has sponsored two reports on computers in schools.

4.15   RESEARCH IN MATHEMATICAL EDUCATION

Throughout this chapter there have been suggestions concerning topics and areas for research in mathematical education. The reader will, however, find many other valuable possibilities outlined in the survey books of Section 4 in Reference 37, **3**, and in

100. R.S. Long *et al.,* 'Research in mathematical education' (*Educational Studies in Mathematics,* **2**, 446-468, 1970)

and

101. *Report of a conference on responsibilities for school mathematics in the 70s* (SMSG, 1971)

Accounts of research, under way and completed, are to be found in the many educational journals, some of which (for example, Reference 5) carry survey articles on research from time to time. An extensive survey of American research appears each year in the September issue of Reference 10. A valuable list of research papers in mathematical education is to be found in

102. *Final report of the SMSG panel on research* (SMSG Newsletter No. 39, 1972)

There have also been two exhaustive surveys of American research:

103. M.N. Suydam, *Annotated compilation of research on secondary school mathematics 1930-1970* (2 vols, US Office of Education, 1972)

104. M.N. Suydam and C.A. Riedesel, *Interpretative study of research and development in elementary school mathematics* (3 vols, US Office of Education, 1969)

The US Educational Resources Information Center (ERIC) publishes an abstracting journal, 105. *Research in Education,* which includes items on mathematical education.

A general source book on research in education is

106. R.M.W. Travers (ed.), *Second handbook of research in teaching* (Rand McNally, 1973)

4.16 ADDRESSES OF ORGANISATIONS

APLET (Association for Programmed Learning and Educational Technology), 33 Queen Anne St., London W1

ATCDE (Association of Teachers in Colleges and Departments of Education), 1 Crawford Place, London W1; now merged with ATTI to form NATFHE (National Association of Teachers in Further and Higher Education), Hamilton House, Mabledon Place, London WC1H 9BH

ATM (Association of Teachers of Mathematics), Market St. Chambers, Nelson, Lancashire, BB9 7LN

BCS (British Computer Society), 29 Portland Place, London W1N 4HU

Centre for Science Education, Chelsea College, Bridges Place, London SW6

CEG (Computer Education Group), North Staffordshire Polytechnic, Blackheath Lane, Stafford

CET (Council for Educational Technology), 3 Devonshire St., London, W1N 2BA

Educational Testing Service, Princeton NS 08540, USA

ERIC (Educational Resources Information Center), Ohio State University, 400 Lincoln Tower, Columbus, Ohio 43210

IFIP (International Federation for Information Processing), POB 6400 Amsterdam

IMA (Institute of Mathematics and its Applications), Maitland House, Warrior Square, Southend-on-Sea, SS1 2JY

MA (Mathematical Association), 259 London Rd, Leicester, LE2 3BE

MAA (Mathematical Association of America), 1225 Connecticut Ave., N.W., Washington, DC 20036

NCC (National Computing Centre), Oxford Rd, Manchester M1 7ED

NCTM (National Council of Teachers of Mathematics), 1906 Association Drive, Reston, Va 22091

NFER (National Foundation for Educational Research), The Mere, Upton Park, Slough, Bucks.

OISE (Ontario Institute for Studies in Education), 102 Bloor St. West, Toronto 5, Ontario

OECD (Organisation for Economic Cooperation and Development), 2 rue André-Pascal, 75 Paris 16^e

The Schools Council, 160 Gt Portland St., London, W1N 6LL

SMP (School Mathematics Project), Westfield College, Kidderpore Ave., London NW3

Unesco (United Nations Educational, Scientific and Cultural Organisation), Place de Fontenoy, 75700 Paris

5

History of Mathematics

I. Grattan-Guinness

5.1 INTRODUCTION

The history of mathematics involves a number of problems in research and even reading which are not normally prominent in mathematical work. To summarise, they are in sequence: (1) there is need to use bibliographies of all kinds to discover what literature (be it mathematical or historical) needs to be read; (2) there may then be difficulty in finding copies of the literature required; (3) unpublished materials may have to be used; and (4) there are important problems of historical interpretation involved in the use of all this literature and in the writing of history.

Space prohibits any kind of exhaustive treatment of these problems here. In the sections below I have taken them in the above sequence, and either indicated principal sources of information available or exemplified the kinds of research carried out or in progress. I have tended to concentrate the detail on the mathematics of the last 200 years (which is likely to be the chief interest of readers), and also on current historical work which by definition may not be described elsewhere. With regard to publications, I have listed only books and journals in or close to the field; I have made no attempt to cover articles published in (those or other) journals. Finally, in order to attune the intention of this chapter to that of the rest of the book, I have borne principally in mind those readers *who take a reasonably serious interest* in historical work rather than bring to it only a passing concern. The information is to date as of Spring 1976.

5.2 BIBLIOGRAPHIES AND CATALOGUES

I shall begin with the principal bibliographical sources for the history of mathematics. They provide *far* more detail than this chapter could hope to equal, and constitute the chief direct sources of information.

K.O. May, *Bibliography and research manual of the history of mathematics* (University of Toronto Press, Toronto/Buffalo, 1973). Over 800 pages in length, this book is the most detailed single bibliographical work available. The five main sections are entitled 'Biography', 'Mathematical topics', 'Epimathematical topics', 'Historical classifications' and 'Information retrieval'. There are also useful hints on methods of storing information collected. The usefulness of the book is rather reduced by the omission of some kinds of information and by many errors of detail.

G. Loria, *Guida allo studia della storia delle matematiche ...* (Hoepli, Milan, 1946). This book ranges well over both problems and information, and contains some very useful bibliographies.

G. Sarton, *The study of the history of mathematics* (Harvard University Press, Cambridge, Mass., 1936: reprinted 1957, Dover, New York). This small volume is particularly useful for its bibliography, especially for details (omitted in May's *Bibliography*) of collected works of mathematicians. The reprint contains also Sarton's companion *The study of the history of science.*

G. Sarton, *Horus: a guide to the history of science* (Chronica Botanica, Waltham, Mass., 1952). A greatly extended version of Sarton's *Science* volume, it contains many points of guidance on the history of all the sciences.

F. Russo, *Eléments de bibliographie de l'histoire des sciences et des techniques* (Hermann, Paris, 1969). A useful and compact volume, it is similar in some ways to Sarton's *Horus,* though not so wide-ranging.

C.C. Gillispie (ed.), *Dictionary of scientific biography* (Scribners, New York, 1970-   ). An ambitious project composed of biographical articles on scientists, it suffers the ailment common to encyclopaedias of stately progress; 'Z' is not yet in sight. The articles on mathematicians seem usually to be at least competent, and the best are excellent.

Isis critical bibliography. Published annually as a supplement to the year's volume of *Isis,* the journal of the History of Science Society (of

America), it contains an excellent survey of current literature in the history of all the sciences.

M. Whitrow (ed.), *Isis cumulative bibliography* (Mansell, London, 1971-   ). Sarton started *Isis* and its critical bibliography in 1913. Mrs Whitrow's monumental project accumulates the bibliographies up to 1965 under major headings ('Personalities', 'Institutions', and so on).

Royal Society of London, Catalogue of scientific papers (19 vols and 4 index vols 1867-1925, Cambridge and London; Cambridge University Press (and others): reprinted 1965, Johnson and Kraus, New York). A comprehensive survey of nineteenth century scientific literature, the index volume for mathematics is fortunately one of those that was completed.

International catalogue of scientific literature, section A, Mathematics (14 vols, Royal Society, London, 1902-1917: reprinted in 1 vol., 1968, Johnson, New York). The successor to the Society's *Catalogue,* it is of comparable style and standard for the early part of this century.

(See Section 3.4 for further details of the last two items.)

In addition to these sources, the ones cited in Chapter 3 also apply, and some comments would be suitable here on their utility for historical work. Of the current abstracting journals, the *Zentralblatt für Mathematik* is the most comprehensive, although the use there of self-reviews reduces the measure of critical writing. *Mathematical Reviews* is astonishingly meagre, while *Poggendorff,* dependent on the strength of inclination of its contributors to autobiography, is patchy. *The Jahrbuch über die Fortschritte der Mathematik* proves an excellent service for the years (1868-1942) that it covered.

5.3   JOURNALS FOR THE HISTORY OF MATHEMATICS

Many of the articles on the history of mathematics appear in a few mathematical journals, several history of science journals, a few obituary journals associated with major scientific institutions, and some journals in mathematical education. But there are or have been journals in the history of mathematics, of which the chief ones are described now in chronological order of inauguration:

Bullettino di Bibliografia e di Storia delle Scienze Matematiche (ed. and published B. Boncompagni, 20 vols, Rome, 1868-1887: reprinted n.d., Johnson, New York). The wealthy editor–publisher had the habit of changing the text after sheet-proofs had been passed, so that copies are not all identical. There is much useful bibliographical information in these volumes as well as some important articles.

Abhandlungen zur Geschichte der Mathematik . . . (ed. M. Cantor, 9 + 21 vols, Teubner, Leipzig, 1877-1913). The first series was published as a supplement to the *Zeitschrift für Mathematik und Physik,* and is usually bound with it.

Bibliotheca Mathematica (ed. G. Eneström, 3 + 13 + 14 vols, Teubner, Stockholm and (later) Leipzig, 1884-1915). The first series was published as a supplement to *Acta Mathematica,* and is usually bound with it. As the years passed, the journal became steadily more of an anti-Bible to M. Cantor's *Vorlesungen über die Geschichte der Mathematik,* correcting mistakes and re-locating all the previous corrections. Many articles are excellent.

Bollettino di Bibliografia e Storia delle Matematiche (ed. G. Loria, 6 + 21 vols, Turin, 1898-1919). The first series was published as an appendix to the *Giornale di Matematiche,* with which it is often bound. After the latter's demise it continued as an appendix in the *Bollettino di Matematica.* It must not be confused with Boncompagni's *Bullettino.*

Quellen und Studien zur Geschichte der Mathematik (ed. O. Neugebauer and others, 4 (*Quellen*) + 4 (*Studien*) vols, Berlin, 1929-1938). Some important lengthy articles appeared in this journal.

Scripta Mathematica (ed. various, Yeshiva University, New York, 1932-   ). Owing to the enthusiasm of R.C. Archibald, the early volumes contained much useful material. Books in a companion series entitled *Scripta mathematica library* were published from time to time.

Bulletin Signalétique, numéro 522. Histoire des Sciences et des Techniques (Centre National de la Recherche Scientifique, Paris, 1941-   ). This series is comparable in style with the series *Mathématiques* mentioned in Section 3.3.

Istoriko-matematicheskogo Issledovaniya (Nauk, Moscow, 1948-   ). This journal contains only articles in Russian, and is the chief (but not the only) source of Russian articles in the history of mathematics.

Archive for History of Exact Sciences (ed. C.A. Truesdell III, Springer, Berlin, 1960-   ). Some articles in this journal are of monograph length.

Philosophia Mathematica (ed. J. Fang, Paideia, (now) Memphis, Tenn., 1964-   ). Many articles have an historical aspect.

Historia Mathematica (ed. K.O. May, University of Toronto Press, Toronto, 1974-   ). Intended as an international journal in the history of mathematics, it contains an extensive abstracting section of its own.

5.4   BOOKS AND EDITIONS

The number of books in this field, though not enormous, is much too
varied or extensive to receive a detailed assessment here. The
bibliographies cited in Section 5.2 should provide at least the factual
element required. In this section I offer a brief outline of significant
books and editions, concentrating on the most recent work. By way of
preface, I mention that the most important publishers of books in the
history of mathematics are several American and English university
presses, and Springer. Olms, Dover, Chelsea and Blanchard have the best
reprint lists, although the bibliographical information appended is
sometimes inadequate.

5.4.1   General histories

Reliance on such volumes for information and enlightenment should be
limited. However eminent the author or substantial his volumes, the
omissions are bound to be significant and the errors perhaps
fundamental (as Eneström noted of Cantor; cf. Section 5.3). Several are
not good. The most impressive of the *genre* is M. Kline's *Mathematical
thought from ancient to modern times* (Oxford University Press, New
York, 1972), but even his 1250 pages are silent on many things. Among
the compact volumes, D.J. Struik's *The concise history of mathematics*
is by far the best; but 'concise' is very much the word, for even the
latest edition (the fifth, in German: VEB Deutscher Verlag der
Wissenschaften, Berlin, 1972) has only one chapter for each century
from the seventeenth on. N. Bourbaki, *Éléments d'Histoire des
Mathématiques* (3rd edn, Hermann, Paris, 1974) is useful for some
developments in this century.

The largest single source of information is still the famous
Encyklopädie der mathematischen Wissenschaften . . . (6 parts, each in
several volumes: Teubner, Leipzig, 1898-1935). The encyclopaedia was
written at the time of the best writing in the history of mathematics.
The amount of mathematical literature covered is staggering, and even
then in the early years further substantial articles appeared as *Berichte*
of the *Jahresbericht der Deutschen Mathematiker-Vereinigung.* But the
standard of historical interpretation is usually slight; even the question
of relevance is rarely discussed. The French began their own expanded
version with the *Encyclopédie des sciences mathématiques . . .* under
the general editorship of J. Molk, but his death in 1914 terminated the
project with no part completed and some left even in mid-article. The
British planned a version, but interest was so slight that nothing was
published at all. However, the Germans began a second edition in the

late 1930s (Teubner Verlagsgesellschaft, Leipzig). Perhaps there is a lesson here about national character.

Another useful general source of information is the various large-scale national and international encyclopaedias, which contain not only biographical but also technical articles. Often these are slight and derivative in character, but sometimes they are substantial. Further, they can be used *as primary literature* to discover what were *thought* to be the most important features of a mathematical discipline at a given time and how the different disciplines were ranged in importance. The famous *Encyclopédie on dictionnaire raisonné des sciences . . .* (28 vols, ed. J. le R. d'Alembert and D. Diderot, Paris and elsewhere, 1751-1765) is very useful in this respect, although its importance as a source of ideas for contemporary scientists is probably exaggerated. The equally famous eleventh edition of the *Encyclopaedia Britannica* (29 vols, ed. H. Chisholm, Cambridge University Press, 1910-1911) is particularly responsive to such treatment.

5.4.2 Source books

The idea of these books is to give the reader access to classic texts in his own language. The idea is good, but the results have been very patchy; only one of my acquaintance (cf. subsection 5.4.9) seems successful. The problem is that usually too much ground is covered, so that no intelligible or insightful impression is gained. D.E. Smith, *Source book in mathematics* (McGraw-Hill, New York, 1929; reprinted in 2 vols, Dover, New York, 1959) is a well-known exemplification of the tendency. Among more recent works even D.J. Struik, *A source book in mathematics 1200-1800* (Harvard University Press, Cambridge, Mass., 1969) exhibits the difficulties.

5.4.3 Collected and selected editions

There are many of these, even for some fairly obscure mathematicians. Several, especially French ones, are reduced in merit by excessive or questionable selection, anachronistic modernisations of notations and insufficient bibliographical information. Many of the principal ones are listed in Sarton's *Guide*, though not in May's *Bibliography* (cf. Section 5.2). Quite a number have been reprinted lately, especially by Chelsea, Olms and Springer. Recently completed new ones include those for Borel, Brouwer, Cauchy (93 years in the making!), Hadamard,

Lebesgue, Schur and Weyl. Among those forthcoming or in progress are, in alphabetical order of mathematician:

Bernoullis *Gesamtausgabe* (ed. Schweizer Naturforschenden Gesellschaft, over 20 vols, Birkhäuser, Basel, 1955-   ). All the publications and manuscripts of the mathematical Bernoullis (and also J. Hermann) are to be included. Three volumes have appeared so far.

Bernard Bolzano Gesamtausgabe (ed. E. Winter and others, over 50 vols, Frommann, Stuttgart, 1969-   ). This edition is to contain all of Bolzano's published work and extant manuscripts, and many volumes will cover his work in mathematics and logic.

Leonhardi Euler opera omnia series quarta. The first three series, containing all of Euler's published writings, are nearing completion from Orell Füssli at Zurich. The fourth series will be the most valuable, since it will contain Euler's large correspondence and many manuscripts mostly previously unpublished (although Russian editions of some manuscripts have been issued). I understand that the project is a joint Swiss–Russian venture. The (Western) publisher is Birkhäuser at Basel. An introductory volume already available is *Descriptio commercii epistolici* (ed. A.P. Yushkevich and others, 1975).

The *Collected works* of Kummer are in course of appearance in two volumes from Springer-Verlag, edited by A. Weil.

While the Leibniz *Sämtliche Briefe und Schriften* grinds on, several important individual studies have been published. See especially H.J. Zacher, *Die Hauptschriften zur Dyadik von G.W. Leibniz* (Klostermann, Frankfurt am Main, 1973); E. Knobloch, *Die mathematischen Studien von G.W. Leibniz zur Kombinatorik* (Steiner, Wiesbaden, 1973) and *Ein Dialog zur Einführung in die Arithmetik und Algebra* (Frommann, Stuttgart, 1976); P. Costabel, *Leibniz and dynamics . . .* (Methuen, London, 1973); E.A. Fellmann (ed.), *Marginalia in Newtoni Principia Mathematica (1687)* (Vrin, Paris, 1973), where Leibniz's annotations to the book are presented; and J.E. Hofmann, *Leibniz in Paris. . .* (trans. A. Prag, Cambridge University Press, 1974).

E.G. Forbes, *The unpublished writings of Tobias Mayer* (3 vols, Vandenhoeck and Ruprecht, Göttingen, 1972-1973), together with his edition of *The Euler-Mayer correspondence* (Macmillan, London, 1971), shed new light on eighteenth century mathematics and astronomy.

The mathematical papers of Isaac Newton (ed. D.T. Whiteside, 8 vols, Cambridge University Press, 1967-   ). This is a complete edition of Newton's manuscripts in mathematics, with interpretative commentary on many related topics. In addition, in 1973 Harvard University Press published a two-volume variorum edition of Newton's *Principia* edited by A. Koyré and I.B. Cohen, together with a companion *Introduction* by Cohen.

Selected writings of Giuseppe Peano (trans. and ed. H.C. Kennedy, University of Toronto Press, Toronto, 1973). A useful selection of translations from Peano's writings is provided, together with a complete bibliography.

The new elements of mathematics of Charles S. Peirce (ed. C. Eisele, 5 vols). To appear from Mouton at the Hague, this edition will contain all Peirce's mathematical (but not his many logical) manuscripts.

W. Sierpinski, *Oeuvres choisies* (ed. S. Hartman and others, 3 vols, Polish Academy of Sciences, Warsaw, 1974). 'Choisies' is the word for it, for Sierpinski published about 800 papers. None of his 50 books is included.

N. Wiener, *Collected works* (ed. P. Masani). This edition has been announced by M.I.T. Press at Cambridge, Mass., who put out a *Selecta* in 1964.

5.4.4   National histories

Up to the Middle Ages mathematics was dominated by the Egyptians, the Babylonians, the Greeks and the Arabs. Then England, France, Switzerland and Germany took most of the limelight. This fact may have subconsciously motivated the writing of national histories, starting with G. Libri's *Histoire des sciences mathématiques en Italie* (4 vols, Renouard, Paris, 1838-1841). A notable recent addition is A.P. Yushkevich, *Istoriya matematiki v Rossii* (Nauk, Moscow, 1968).

5.4.5   Educational and institutional history

Such volumes are far too infrequent and often disappointing, but a magnificent exception is K.R. Biermann, *Die Mathematik und ihre Dozenten an der Berliner Universität 1810-1920* . . . (Akademie-Verlag, Berlin, 1973).

5.4.6   Correspondence

Again, there are all too few editions. Rather disappointing is H. Minkowskii, *Briefe an David Hilbert* (ed. L. Rüdenberg and H. Zassenhaus, Springer, Berlin, 1973), for only one side of the correspondence is available and that edited in ignorance of scholarly principles. Better is *Pic'ma Karla Weierstrassa k Sofe Kovalevsky* (ed. P. Ya. Polubarinova-Kochina, Nauk, Moscow, 1973), containing an edition in Russian and German of the surviving exchanges. The famous public correspondence between Ferrari and Tartaglia on the solution of cubic equations is at last easily available in A. Masotti (ed.), *Cartelli di sfida matematica* (Ateneo di Brescia, Brescia, 1974).

5.4.7 Biographies and studies

Recent works include C.C. Gillispie, *Lazare Carnot savant* (Princeton University Press, Princeton, N.J. 1971); C. Reid, *Hilbert (Springer,* Berlin, 1970), better than its journalistic American would imply; M.S. Mahoney, *The mathematical career of Pierre de Fermat 1601-1665* (Princeton University Press, Princeton, N.J., 1973); my and J.R. Ravetz's *Joseph Fourier 1768-1830* (M.I.T. Press, Cambridge, Mass., 1972); H. Wussing, *C.F. Gauss* (Teubner, Leipzig, 1974), the best introductory volume on Gauss yet available; J. Shirley (ed.), *Thomas Harriot, Renaissance scientist* (Clarendon Press, Oxford, 1974), a survey of many of Harriot's achievements in prelude to a projected edition of his manuscripts; and P. Dugac, *Richard Dedekind et les fondements des mathématiques* (Vrin, Paris, 1976). Under this heading may also be put P.L. Rose, *The Italian Renaissance of mathematics. Studies on humanists and mathematicians from Petrarch to Galileo* (Droz, Geneva, 1975).

5.4.8 Ancient mathematics

I shall mention only three recent works in this always most popular area of historical research. A. Szabo, *Anfänge der griechischen Mathematik* (Oldenburg, Munich, 1969) is a significant volume; and R.J. Gillings, *Mathematics in the time of the Pharaohs* (M.I.T. Press, Cambridge, Mass., 1972) and W.R. Knorr, *The evolution of the Euclidean elements* (Reidel, Dordrecht, 1975) have aroused interest.

5.4.9 Logic and set theory

The developments of the last 100 years are of considerable interest because of their enduring philosophical implications. J. van Heijenoort (ed.), *From Frege to Gödel* ... (Harvard University Press, Cambridge, Mass., 1967) is a model for its 'source book' series and contains translations and editions of many key papers. N.I. Styazhkin, *History of mathematical logic from Leibniz to Peano* (M.I.T. Press, Cambridge, Mass., 1970), is a translation from the Russian of a moderate account of the subject matter. My *Dear Russell–dear Jourdain* (forthcoming) contains much new information on the writing of *Principia Mathematica*. Medieval logic is well studied now, following the example set in W. Risse, *Die Logik der Neuzeit* (2 vols, Frommann, Stuttgart, 1964-1970); among more recent studies are E.J. Ashworth, *Language and logic in the post medieval period* (Reidel, Dordrecht, 1974), and J. Pinborg, *Logik und Semantik im Mittelalter. Ein Überblick* (Frommann, Stuttgart, 1974). Philosophy in an unusual and interesting sense is treated in G. Canguilhem (ed.), *La mathématisation des doctrines informes* (Hermann, Paris, 1972), a collection of articles, including several historical ones, on how mathematics was and can be brought into new areas.

5.4.10 Number and number theory

L.E. Dickson, *History of the theory of numbers* (3 vols, Carnegie Institution, Washington D.C., 1919-1923, reprinted 1952, Chelsea New York), is still the major source of information. O. Ore, *Number theory and its history* (McGraw-Hill, New York, 1948) is a more compact version of the *genre*. There are several books on the concept of number: recent ones of interest are K. Menninger, *Number words and number symbols* (M.I.T. Press, Cambridge, Mass., 1970), and two from the Bibliographisches Institut in Mannheim: C.J. Scriba, *The concept of number* (1968), and H. Gericke, *Geschichte des Zahlbegriffs* (1970). See also G. Guitel, *Histoire comparée des numérations écrites* (Flammarion, Paris, 1975).

5.4.11 Algebra

Linear algebra is still largely untouched, but abstract algebra has received some treatment in H. Wussing, *Die Genesis des abstrakten Gruppenbegriffes* (Akademie-Verlag, Berlin, 1969), and L. Novy, *Origins of modern algebra* (Czechoslovak Academy of Sciences, Prague, 1973).

5.4.12 Calculus and mathematical analysis

In addition to the works of Whiteside and Mahoney cited in subsections 5.4.3. and 5.4.7, other recent books include T. Hawkins, *Lebesgue's theory of integration . . .* (University of Wisconsin Press, Madison, Wisc., 1970); I.N. Pesin, *Classical and modern integration theories* (Academic Press, New York, 1970), originally in Russian, similar but much inferior to Hawkins's book; I. Grattan-Guinness, *The development of the foundations of mathematical analysis from Euler to Riemann* (M.I.T. Press, Cambridge, Mass., 1970); D. van Dalen and A.F. Monna, *Sets and integration*... (Wolters Noordhoff, Groningen, 1972); G. D. Birkhoff (ed.), *A source book in classical analysis* (Harvard University Press, Cambridge, Mass., 1973), containing translations (where necessary) of bits of classical papers; and F.A. Medvedev, *Razvitie ponyatiya integral* (Nauk, Moscow, 1974). All these works concentrate on foundational aspects; applications have not earned the same interest, but see A.F. Monna, *Functional analysis in historical perspective* (1973) and *Dirichlet's principle* ... (1975), both from Oosthoek, Scheltema and Holkema at Utrecht.

5.4.13   Geometry

The subject of geometry has been utterly transformed since 1800, and yet the corresponding history so ignored that old texts such as R. Bonola, *Non-Euclidean geometry* (especially with the additions in the reprint of 1955, Dover, New York), and D.M.V. Sommerville, *Bibliography of non-Euclidean geometry* ... (Harrison, London, 1911), are still basic. Kline's volume listed in subsection 5.4.1 is strong in this area, but we still await especially the history that will reduce non-Euclidean geometry to its proper status in nineteenth century mathematics.

5.4.14   Mathematical physics (including mechanics)

As with geometry, this area of great importance is sadly neglected. Here the unsurpassed source of data is H. Burkhardt's 'Entwicklung nach oscillierenden Funktionen ...', an 1800-page supplement (1900-1908) to Vol. 10 of the *Jahresbericht der Deutschen Mathematiker-Vereinigung*. Among more recent work, C.A. Truesdell's *The rational mechanics of flexible or elastic bodies 1638-1788* (Orell Füssli, Zurich, 1960), a volume in Euler's *Opera omnia* (cf. subsection 5.4.3), and his *Essays in the history of mechanics* (Springer, Berlin, 1968) are essential reading, even though insight battles with idiosyncrasy throughout.

5.4.15 Astronomy and cosmology

This is a lively area in the history of science, although of course only part of the work is mathematical in content. The most comprehensive reference to give is to the *Journal for the History of Astronomy* (ed.) M.A. Hoskin, Science History Publications, (now) Cambridge, 1970-) . Among recent books, Kepler was fairly thoroughly studied in *Internationales Kepler-Symposium, Weil der Stadt 1971* (Gerstenberg, Hildesheim, 1973). A mass of publications appeared in 1973 for the Copernicus quincentenary; among the valuable fragments there is an incredible bargain at $1 of Vol. 117 (1973), no. 6 of the *Proceedings of the American Philosophical Society.* Among non-commemorative volumes, E.J. Aiton, *The vortex theory of planetary motions* (Macdonald, London, 1972), has been acclaimed. Springer have published O. Neugebauer's three-volume *A history of mathematical astronomy* to inaugurate their two new *Studies* and *Sources* series in the history of mathematics and physical sciences.

5.4.16 Probability and statistics

Some feeling for the achievements since 1800 may be gained from E.S. Pearson and M.G. Kendall (eds.), *Studies in the history of probability and statistics* (Griffin, London, 1971), a volume of articles from the journal *Biometrika.* On aspects of earlier developments, try N.L. Rabinovich, *Probability and statistical inference in ancient and medieval Jewish literature* (University of Toronto Press, Toronto, 1973); or R. Rashed (ed.), *Condorcet. Mathématiques et société* (Hermann, Paris, 1974), a selection of texts with commentary. Otherwise, once again the territory is largely virgin.

5.4.17 Computing

Surprisingly much has been done in this comparatively new area. See C. and R. Eames, *A computer perspective* (Harvard University Press, Cambridge, Mass., 1973); H.H. Goldstine, *The computer from Pascal to von Neumann* (Princeton University Press, Princeton, N.J., 1973); and B. Randell (ed.), *The origins of digital computers: selected papers* (Springer, New York, 1973). But numerical analysis, now closely associated with computing, has its much longer history almost ignored.

5.5   LIBRARIES AND CATALOGUES

So far I have concentrated on the first problem of Section 5.1, namely, the search for references and bibliography. In this section I shall touch on the second problem—that of obtaining the literature. Sometimes the usual methods of library catalogues and even interlibrary loan fail because of the rarity of the material. It is worth knowing of special collections of books and journals, and I shall now describe some typical ones. Others are mentioned from time to time in the 'Sources' department which was started by the author in *Historia Mathematica* (cf. Section 5.3).

Institut Mittag-Leffler, S 182-62 Djursholm, Auravägen 17, Sweden. Mittag-Leffler's old home, the library is based on his own collection and contains a magnificent range of volumes, especially rich in nineteenth century mathematics*.

Graves library, University College, London WC1, England. J.H. Graves was Professor of Jurisprudence there from 1838 to 1843, and a wealthy amateur mathematician. His library is particularly rich in nineteenth century volumes and contains many very rare items†.

De Morgan library, Senate House, University of London, London WC1, England. A. de Morgan's rich collection of books and offprints contains several with interesting annotations from the authors. In connection with it consult his *Arithmetical books . . .* (London, 1847: reprinted with D.E. Smith, *Rara arithmetica,* Chelsea, New York, 1970).

David Eugene Smith Library, Columbia University, New York, N.Y.10027, USA. This enormous collection of volumes was left to the university by an eminent historian of mathematics.

Brown University, Providence, Rhode Island 02912, USA. This university has one of the very best mathematical libraries, and also a department for the history of mathematics which specialises in ancient mathematics.

In addition to these and other special collections, many national and institutional libraries are extremely extensive. However, their catalogues are not always straightforward to use. (The same can be true of the collections mentioned above!) For example, it is important to know that, *pace* the instruction in the *Catalogue* of the British Museum to look up the name of the journal under the title and then pursue the

---

* See my 'Materials for the history of mathematics in the Institut Mittag-Leffler', *Isis,* **62,** 363-374 (1971); and St. Grönfeldt, *G. Mittag-Lefflers matematiska bibliothek,* Djursholm, Stockholm (1914).

† See A.R. Dorling, 'The Graves mathematical collection in University College London', *Annals of Science,* **33,** 307-310 (1976).

cross-reference, many of the journals held there are *not* so listed. They are to be found under some sub-entry for an appropriate country, city or institution, and guess-work will have to get you there.

5.6 MANUSCRIPTS AND ARCHIVES

This section takes up the third problem of Section 5.1, namely the use of unpublished materials. There are a number of large-scale projects on hand in the history of science, of which some take mathematics into account. In addition, there are several national archives (the British Library Reference Division, for example) which hold substantial collections of material. Assuming the latter to be too well-known (or at least easily traceable) to require further elaboration here, I shall repeat the scheme of the last two sections and describe a representative sample of specific projects and institutions of particular importance. Once again the 'Sources' department of *Historia Mathematica* (cf. Section 5.3) will come in useful for the interested reader, as also will 'Projects in progress', another department which was started there by the author.

R.M. MacLeod and J.R. Friday, *Archives of British men of science* (Mansell, London, 1972). Published on microfiche with an accompanying booklet, this list describes the papers of a selection of prominent British scientists. The amount of detail is very uneven and so far mathematicians have not been much dealt with, but apparently future editions may take them more into account.

Center for History and Philosophy of Physics, 335, E 45 St, New York, N.Y.10017, USA. The Center was instituted by the American Institute of Physics in 1965 both to store and catalogue the papers of eminent physicists and also to list the manuscripts of all kinds held in other American institutions. Since the Center includes mathematical physicists and astronomers within its compass, it has information on quite a number of mathematicians. There ought to be a comparable institution for mathematics, but I imagine that the chances of it are minimal.

The Contemporary Scientific Archives Centre was recently set up at Oxford University by Professor M. Gowing with the co-operation of the Historical Manuscripts Commission. The Centre catalogues the papers of recent scientists of all kinds (including mathematicians) and then deposits them in the archives of a suitable institution.

Deutsches Zentralarchiv, Historische Abteilung II, 48, Weisse Mauer, Merseberg, East Germany. This is a magnificent collection of Prussian archivilia, specialising in state papers for all Prussian universities in the nineteenth century. It is essential for any biographical or institutional history of nineteenth century German mathematics.

École Nationale des Ponts et Chaussées, 28, rue des Saints-Pères, Paris 7, France. The library is an excellent example of manuscript materials held by the French écoles. See *Catalogue des manuscrits de la bibliothèque de l'École Nationale des Ponts et Chaussées* (Paris, 1886).

Institut Mittag-Leffler (cf. Section 5.5). As well as a remarkably fine library, the Institut contains many manuscripts of great importance.

The National Aeronautics and Space Administration, Washington, D.C. 20546, USA set up an Historical Office in 1962 to co-ordinate the collection of manuscript materials of all kinds in connection with its activities. As well as producing a number of publications, the Office organises a seminar every summer.

Niedersächsische Staats- und Universitäts-bibliothek, Handschriften-abteilung, 34 Göttingen, Prinzenstr. 1, West Germany. An outstanding example of the materials held by a German university, it contains the papers of Gauss, Dedekind, Hilbert, Hurwitz (correspondence only) and Klein, as well as many other smaller collections of importance to the history of mathematics.

A co-ordination system facilitates the discovery of materials in West German archives. Scholars requiring information should write to: Verzeichnis der Nachlässe in deutschen Archiven, 3500 Kassel, Brüder-Grimm-Platz 4A, West Germany. Some information can be recovered from volumes in the series *Verzeichnis der schriftlichen Nachlässe in deutschen Archiven und Bibliotheken,* published by Boldt at Boppard am Rhein.

5.7   SOCIETIES

There are a number of mathematical and history of science societies which concern themselves, if very slightly, with the history of mathematics. But there are also two societies for the subject:

British Society for the History of Mathematics. Secretary: Mr J.J. Gray, The Open University, Milton Keynes, Bucks., England.

Canadian Society for the History and Philosophy of Mathematics. Secretary: Dr J.L. Berggren, Department of Mathematics, Simon Fraser University, Burnaby, B.C., Canada.

In addition to formal societies, there is the informal world grapevine system, which usually functions well in this subject. The best current means of discovering historians of mathematics and their interests is to consult K.O. May, *World directory of historians of mathematics* (Historia Mathematica, Toronto, 1972), of which a second edition is in preparation.

5.8   THE OPEN UNIVERSITY

In the United Kingdom the Open University has introduced a half-unit in the history of mathematics. There is a general course using eleven television programmes and presupposing no initial knowledge on the part of the students, and two special courses based on some radio programmes and also on specially written booklets. One course deals with the development of the calculus up to the early nineteenth century and the other with the concept of number.

5.9   STATE AND NATURE OF THE ART

So far this chapter has been given over to the supply of information required to answer the first three problems posed in Section 5.1. This section is devoted to the fourth problem, namely the question of interpretation involved in doing historical work or in reading others' historical writings. Again space permits only the exemplification of a few principal points.

I trust that the impression has been conveyed so far that more work has been done, or is in progress, than might have been anticipated. But the fact is undeniable that there are still large and quite fundamental gaps in our knowledge. Many leading figures lack a biographer, several of the major developments in mathematics are still unchronicled, and the history of mathematical education is almost everywhere non-existent. Thus the interested reader will very likely have to resort to do-it-yourself, even if only to the reading of a little primary literature, and it is at this stage that points of history are likely to arise. Put very simply, the history of mathematics contains both history and mathematics, for historical issues arise along with the mathematical ones. It is such issues that I wish to emphasize here*.

The first one concerns purely epistemological questions, mathematics regarded as knowledge independently of its historial genesis. Old work probably belongs to some general area of mathematics which is

---

* As will be clear, I am stating my own views here. I have pursued them in more detail in my 'Not from nowhere. History and philosophy behind mathematical education', *International Journal of Mathematical Education in Science and Technology,* **4**, 321-353 (1973).

still in progress when we read it, or at least which has developed beyond the state that that work manifests. Even the demise of a branch of mathematics is usually some transmogrification into an alternative form. Now the intermediate developments play an important role in our reading of the old work, for they took place after the period we study but before our own time. Thus they are likely to involve the introduction of fresh distinctions, the establishment of new inter-connections, generalisations of various kinds, correction of mistakes, and so on. From the epistemological point of view these developments are very important and desirable; but for historical purposes they are dangerous avenues of *post hoc* verification of old work *as* the (inevitable) progress towards some currently fashionable state. Histori-cal work requires the construction of *ignorance situations* for our historical figures, based on the relevant intermediate developments of which they were essentially ignorant. Ignorance situations are difficult to specify, since by definition mathematicians rarely say much about the things of which they are basically unaware. One way of approximating to an ignorance situation is to examine the research carried out in the intervening period and discover its claimed novelties.

The second historical point is a particularly important case of the first, and concerns *priority structures* within mathematics. Theories change sometimes simply in the status that they do or do not enjoy among mathematicians. The curve of fortune tends to be uninodular: after genesis a theory becomes established and even systematised, and acquires high fashion before passing towards oblivion or replacement by superior alternatives. Now such demises or transformations *are themselves part of the ignorance situation*; and if care is not taken, then the current priority structure will be transplanted *en bloc* into the past under study. In particular, it may be thought that if a theory is dead now, then its history is unimportant today. But this is *not* necessarily so: during its reign the theory may have exercised considerable influence on other theories which now are in full bloom, and in any case must be granted the importance that it had at the appropriate time. The reverse danger also applies. It is very tempting to seek in the past anticipations of currently important mathematics; and if ye seek, then ye shall find. But the similarities may well be superficial, while the differences of conceptions are much more profound though more innate.

Both these historical points emphasise the difference between historical and mathematical evaluations of mathematics. But *no* invalidity is thereby implied for mathematical evaluations themselves, for they may well be the mathematician's only (and quite legitimate) interest; the danger warned against is the danger of misidentification.

The final point is the relevance of historical work to mathematical research. Blanket condemnation of the current anti-historical attitude among mathematicians stoops to the level of polemic, and is too simplistic anyway. Let it suffice to remark that mathematicians always work as part of a historical process; thus they need to know their inheritance as an *active* ingredient of their work, and not as mere anecdotal froth. It is noteworthy that the greatest mathematicans have often had a considerable interest in the history of their subject, and some have made significant contributions to it. For they know that, while there is a distinction between history and epistemology, there is *no* genuine distinction between current mathematical work and its historical past; *there are only mathematical problems and their history.* If we want to learn from our betters, then this is one of the lessons to acquire.

6

Logic and Foundations

W.A. Hodges and G.T. Kneebone

6.1   INTRODUCTION

Mathematical logic is the mathematical study of the main activities of mathematicians: constructing, defining, proving, computing. There are two radically different approaches to this subject. According to the first of these, mathematical logic has the task of *justifying* the activities of mathematicians—without logic, mathematicians might slip into conceptual muddles or even downright contradictions. This first approach is sometimes loosely called the 'foundational' approach, and it was very much to the fore until about 1930. In the 1930s a number of powerful logical techniques were invented, and a different view of logic became popular. According to this second way of thinking, which is sometimes called the 'technical' approach, logic is simply a branch of mathematics on a par with any other—the only distinguishing feature of logic being that it talks about definability, meanings, etc. The foundational kind of research has made relatively little progress in the last 40 years, except among those mathematicians who reject parts of classical mathematics (cf. Section 6.4).

For a remarkably thorough modern textbook, covering the whole field but without a bibliography, see Shoenfield[1]. Enderton[2] and Mendelson[3] are excellent introductory textbooks, while Crossley[4] is a readable short introduction to the field. Barwise[5] is a summary of the main results, including recent research.

There are a number of journals devoted specifically to mathematical logic and related fields, in particular:

Annals of Mathematical Logic (North-Holland, Amsterdam, 1970-   )
Archiv für Mathematische Logik und Grundlagenforschung (Kohlhammer, Stuttgart, 1950-   )
Fundamenta Mathematicae (Polska Akademia Nauk, Warsaw, 1920-   )
The Journal of Symbolic Logic (Association for Symbolic Logic, Princeton and Los Angeles, 1936-   )
Zeitschrift für Mathematische Logik und Grundlagen der Mathematik (Deutscher Verlag der Wissenschaften, Berlin, 1955-   )

The Journal of Symbolic Logic, the journal of the Association for Symbolic Logic, has in the past been particularly useful for its reviews of all books and papers in the field of mathematical logic, but its review section is now being curtailed. *Fundamenta Mathematicae* is devoted mainly to logic and topology.

Every year a number of conferences in logic are held; usually the proceedings are published in the North-Holland series *Studies in Logic and the Foundations of Mathematics,* or in the Springer-Verlag series *Lecture Notes in Mathematics.*

6.2   JUSTIFICATION OF MATHEMATICS

We begin with the main foundational question, which carries all others in its wake: How can mathematics be justified? This question has been asked, and answers have been proposed, both by philosophers interested in mathematics and by philosophically minded mathematicians. There are three excellent collections of readings on the topic: Benacerraf and Putnam[6] is mainly philosophical; van Heijenoort[7] contains the classic contributions from the mathematical side; and Hintikka[8] is a judicious collection of more recent mathematical papers with philosophical implications. A historical survey which concentrates on the period 1899-1931 is to be found in Kneebone[9]; Mostowski[10] is more advanced, and covers 1930-1964.

From a philosophical point of view, we find the main battle-lines drawn between those logicians who take mathematical statements to be literally true and those who do not. Those who take them as literally true are called *Platonists.* A Platonist believes, for example, that the collection of all real numbers is simply there, and simply has whatever cardinality it does have. On this view, a mathematician never produces anything that was not there already; he merely makes explicit a proof or a structure which existed before he defined it. For a Platonist, the main foundational task is to lay down principles which tell us what mathematical structures do exist. It is usually held that the axioms of set theory do this well enough for practical purposes, although there is always a chance of finding useful new axioms (cf. Section 6.7). See the papers of Gödel and Bernays in Reference 6.

Opposed to the Platonists are the *formalists*. They hold that mathematical statements do not literally say anything at all, but are simply strings of symbols derivable in certain formal systems (cf. Section 6.3). On this view, too, mathematicians never really construct the abstract objects that they seem to talk about, since these objects are no more than a grammatical hallucination. Many contemporary logicians adopt a formalist view because they do not want to be bothered with philosophical questions; and this is perhaps one reason why formalists have shown so little interest in explaining or justifying informal mathematical reasoning, although they use it freely. For a defence of formalism, see Cohen's contribution to Scott[11]. For a different view, see Kreisel in References 11 and 8.

In between the extremes of Platonism and formalism lies *constructivism,* the view that mathematical objects have to be constructed in order to exist. This view tends to lead to unorthodox mathematics—we have therefore isolated it in Section 6.4.

6.3 FORMAL THEORIES AND METAMATHEMATICS

Study of the foundations of mathematics, originally a preserve of philosophers, has long since been taken over by mathematicians. The transition began with Frege's *Grundgesetze der Arithmetik* in 1893, and came to fruition in the work of Hilbert and his many collaborators, who perfected classical metamathematics (proof theory), i.e. the study by mathematical means of entire mathematical theories. For theories to be suitable for such study they must be well-defined, and the technique that has been developed for ensuring this is *formalisation.* Since the first edition of Hilbert and Ackermann[12] in 1928, it has been common practice to single out two most basic types of formal calculus, the *first-order predicate calculus with identity* and *higher-order predicate calculus.* The first-order theory is the most commonly used. It has various forms, all essentially equivalent: Hilbert's axiomatic style (see Shoenfield[1] or Enderton[2]), natural deduction (Kalish and Montague[13] present a practical version), the sequent calculus (Lyndon[14]) and semantic tableaux (Beth[15]). Beth[15] describes some proofs of equivalence between different calculi. For higher-order calculi, Hilbert and Ackermann[12] is still a good reference, although Church[16] is fuller.

To formalise a theory within first-order predicate calculus, one adjoins additional symbols and axioms; see Kalish and Montague[13] for some clear and detailed examples. Of the countless formal theories obtained in this way, two in particular have been studied intensively from foundational motives: formal Peano arithmetic and formal set theory. Since most contemporary mathematics can be developed within these two theories (in well-known ways), any justification of these theories will be a major step towards justifying mathematics. 'Hilbert's programme' accordingly aimed to justify mathematics by finding a finitary proof of the consistency of arithmetic. This programme lost all plausibility when Gödel[17] revealed the limitations inherent in formalisation and finitary mathematics. But the phase of metamathematical research which ended with Gödel's paper in 1931 had produced a wealth of techniques and results, and these are discussed fully in the invaluable volumes of Hilbert and Bernays[18].

After Gödel's paper, metamathematicians concentrated their energies on finding normal forms for proofs, and methods for converting proofs to normal form. Unquestionably the most useful normal form has been the *cut-free sequent proofs* of Gerhardt Gentzen; see Kleene's learned book[19] for a good account. Turned upside down, these proofs are formally identical with the semantic tableaux of Beth. Smullyan[20] elegantly presents a form of semantic tableaux, and shows how they can be used to prove various metamathematical results. See Prawitz[21] for normalisation in the natural deduction calculus.

Hilbert's programme itself survived, but in a revised form; instead of looking for a finitary proof of the consistency of arithmetic, one tried to measure exactly how much non-finitary equipment was needed for a proof. Chapter 8 of Shoenfield[1] is a good introduction to this area of research; Schütte[22] is more detailed. For more recent trends, see Kreisel's survey, parts I[23] and II (in Reference 24).

Meanwhile a new wind was blowing from the east. Polish scholars, notably Tarski and Lindenbaum, were finding that first-order calculi could profitably be studied *semantically*, i.e. through their interpretations. This approach led rapidly to highly elegant and highly non-finitary Boolean methods such as are summarised in Rasiowa and Sikorski[25]. These ideas soon blossomed into model theory (Section 6.9). Lyndon[14] makes a neat combination of semantic and proof-theoretic methods.

A formal logical device that has attracted some attention in recent years is the ϵ-symbol, or selection operator, that was first used in the 1920s by Hilbert. For a systematic account of this, see Asser[26].

6.4   INTUITIONISM AND CONSTRUCTIVISM

Intuitionism, as propounded by Brouwer from 1920 onwards, pushed
to its ultimate conclusion the anti-metaphysical attitude towards
mathematics that had already commended itself to such mathematicians as Kronecker, Poincaré and Weyl. Brouwer maintained that
'mathematics is a free creation, independent of experience, which
develops from a single primordial intuition', coming into being in the
consciousness of the individual mathematician, and prior alike to
language and to logic. Starting out from this position, Brouwer
produced a body of intuitionist mathematics, radically different from
'classical' mathematics, and not governed by 'classical' logic. The logical
structure implicit in Brouwer's intuitionism was later codified by
Heyting, who also contributed to the further development of
intuitionist mathematics itself. A key notion in such mathematics is
that of a *choice sequence*, i.e. an indefinitely proceeding sequence of
numbers, where the successive choices can be wholly free or can be
subject to restrictions of various kinds. Research on choice sequences
has been continued recently by Myhill, Kreisel and Troelstra.

For a simple account of the philosophy of intuitionism and the
earlier development of intuitionist mathematics, see Heyting[27]; and for
more recent work see Section 2 of van Rootselaar and Staal[28] and
Sections A–D of Kino, Myhill and Vesley[29].

Less radical than intuitionism, but more or less akin to it, are the
various brands of *constructivism,* a common presupposition of which is
that mathematical objects have to be constructed in order to exist.
Constructivists differ according to the types of construction they allow.
Also there are logicians who, while not themselves constructivists, are
interested to see what can be done with limited methods. Thus
predicative methods use only structures which can be defined 'from
below'. See, for example, Feferman in Reference 8, who allows
induction up to previously defined ordinals only.

Several attempts have been made to recast ordinary mathematics in
constructive form, notably by Bishop in Reference 30 and by Lorenzen
in Reference 31. Constructivism is closely bound up with recursion
theory, to which we now turn.

6.5   RECURSION THEORY

Recursion theory is concerned with characterising formally the intuitive
notion of *effectiveness* of an operation or a process, i.e. its being such
that, in principle at least, it can be carried through mechanically so as
to yield a definite result. (It is clear why constructivists should be

interested in recursion theory.) This theory has evolved, from simple concrete beginnings, into a highly abstract theory of remarkable generality. Effectiveness was first studied in the context of arithmetic (effective computability of a number, definability of a function, and so on), but it was soon universalised with the aid of characteristic functions, Gödel numbering, etc.

The story begins with the Dedekind–Peano conception of the arithmetic of the natural numbers, grounded in numeral induction and primitive recursion. Primitive recursive functions are effectively computable, of course; but one can easily define effectively computable functions which are not primitive recursive. In 1936 Alonzo Church proposed that the *general recursive* functions introduced by Kleene should be adopted as the appropriate mathematical idealisation of the effectively computable functions of natural numbers. This proposal, which is known as *Church's thesis*, has been almost universally accepted. (See Davis[32], pp. 10ff, for a brief discussion.) When computability is made precise in this way, the existence of functions which are well-defined but not effectively computable readily follows.

A veritable encyclopaedia of the entire field of recursion theory, with an extensive bibliography, is Rogers[33]. Péter[34] is a general text, good on primitive recursion, and Goodstein[35] offers a more formal treatment. Smullyan[36] deserves mention as a beautifully written introduction.

Within recent recursion theory one can distinguish the following specialised areas: (a) computability theory, (b) degrees of unsolvability, (c) generalised recursion theory and hierarchies, (d) recursive analogues of classical mathematics, (e) decision problems. For (a)–(d), see below; for (e), see Section 6.6.

6.5.1 Computability theory

Several mathematical definitions of 'computation' have been proposed. See Davis[32] or Hermes[37] for good accounts of Turing machines, Kleene's general recursive functions and Post production systems; Mendelson[3] for Markov algorithms; and Shepherdson and Sturgis[38] for register machines. The same functions are computable on all these definitions; proofs are given in the above references, Reference 32 in particular. The collection edited by Davis[39] is a handy source for the early work of Gödel, Church, Kleene, Post and Turing in this field.

It is natural also to ask what can be computed by various other types of machine. On finite automata, pushdown automata, stochastic automata and self-replicating automata, Arbib[40] is full and clear. Some of these machines have close links with mathematical language theory, such as context-free languages (Ginsburg[41]) or regular languages (Conway[42]); see Gross and Lentin[43].

The theory of *program schemes* applies recursion theory to the study of computer programs; see Manna[44]. An earlier attempt to formalise schemes of computation was Church's *lambda-calculus*; see, for example, Hindley, Lercher and Seldin[45] and the references given there.

The journal *Information and Control* (Academic Press, New York, 1958-   ) publishes papers on computability theory.

6.5.2   Degrees of unsolvability (otherwise known as *Turing degrees*)

One may ask what functions could be computed by a Turing machine fitted with an 'oracle' to tell it the values of some non-effective function. The resulting theory is extremely abstract and combinatorial, but in the last 25 years it has exercised some of the best minds in logic and stimulated work in other areas. So-called *priority* methods belong here. See Shoenfield[46] or Yates[47].

6.5.3   Generalised recursion theory and hierarchies

Several approaches to recursion theory—notably arithmetic definability and equation schemes—suggest natural generalisations. We may, for example, consider sets of numbers defined by inductive definitions, or we may replace the set of natural numbers by some other ordinal, or indeed by an arbitrary structure. It turns out that several apparently quite different generalisations lead to very much the same place in the end. Higher-type recursion theory, hyperarithmetic sets and the theory of admissible ordinals (Barwise[48]) all belong here. Inductive definitions have proved to be a powerful unifying principle; see Moschovakis[49] and the references he gives. Fenstad and Hinman's collection[50] presents a variety of approaches, together with a substantial bibliography.

Descriptive set theory (cf. Section 6.8) can also be viewed as a form of generalised recursion theory.

6.5.4   Recursive analogues

There have been a number of studies of the analogues of classical mathematical structures within recursion theory; cf. constructivism

(Section 6.4). For examples, see Mazur[51] on analysis, and Crossley and Nerode[52] on recursive homomorphisms between structures.

6.6 DECISION PROBLEMS

A *decision problem* consists in trying to find an effective procedure which will decide which things possess some specified meta-mathematical characteristic.

In the metatheory of pure logical calculi we have a decision problem of a very obvious kind: to find, for a given calculus and a given class of formulae, a procedure that will determine in a finite number of steps whether any particular formula of the class is or is not formally derivable. In some instances the problem has been solved positively, by production of a suitable procedure; in some others it has been solved negatively, by a proof that no such procedure can exist. Negative solutions of decision problems, particularly for mathematical theories, tend to rely heavily on recursion theory (cf. Section 6.5).

An early instance of a mathematical decision problem was Hilbert's Tenth Problem (1900): to find a method for determining whether any given diophantine equation has a solution. This famous problem, long unsolved, was at last solved negatively in 1970 by Matijasevič; see Davis[53] or Matijasevič himself in Suppes *et al.*[54]

Another challenging decision problem was the word problem for groups (Dehn, 1911): given any finite presentation for a group, to find a method for determining whether any particular word represents the group identity. This problem was solved, also negatively, in 1954 by P. Novikov. For an excellent account of this result, see Rotman[55]. Boone, Cannonito and Lyndon[56] and Miller[57] discuss related problems, such as the conjugacy and isomorphism problems for groups.

When first-order theories (cf. Section 6.3) became an object of general study, it was natural to ask: given a first-order theory *T*, is there an effective method for determining which first-order sentences are consequences of *T*? The answer is known for many important theories *T*; see Ershov *et al.*[58] for a list together with references. Gödel's theorem is the basis for most of the negative answers; see Tarski, Mostowski and Robinson[59] and the later and more powerful method of Rabin (in Reference 60). Ackermann[61] surveys the positive results for pure first-order logic (where *T* is taken to be empty). Two interesting recent positive solutions are for the theory of finite fields (Ax[62]) and the theory of *p*-adic fields (Cohen[63]).

Most non-trivial decision problems in logic have negative solutions; so it was surprising when Rabin[64] proved that the second-order monadic theory of two successor functions was effectively decidable. Rabin's proof is difficult; Siefkes[65] is of some help.

6.7   AXIOMATIC SET THEORY

Axiomatic set theory began in 1908, when Zermelo proposed a system of axioms as an answer to the challenge which the paradoxes had posed to the soundness of Cantor's 'naïve' treatment of sets. Subsequent incorporation of a number of additions and modifications produced the standard Zermelo–Fraenkel system ZF. A somewhat different axiomatic approach was proposed in 1925 by von Neumann; and this is the original ancestor of the current Neumann–Bernays–Gödel system NBG. Yet another system favoured today by many mathematicians is that of Morse and Kelley, MK. In addition, we have a system proposed by Ackermann (see Reinhardt[66]) which is essentially equivalent to ZF but based on different insights. There is also the personal system of Quine, presented in Reference 67 with trenchant philosophical remarks along the way.

For a readable and comparatively informal account of the ZF axioms, their relation to the rest of mathematics, and how they avoid the paradoxes, see Fraenkel, Bar-Hillel and Levy[68]; and for a succinct account of the history of the set-theoretic axioms, see Fraenkel's introductory section in Bernays and Fraenkel[69]. An easy-going treatment of ZF set theory, up to the handling of ordinal and cardinal numbers, is to be found in Suppes[70]; and there is a tighter presentation of the theory, based this time on MK, in Monk's admirable textbook[71]. Takeuti and Zaring[72] is very formal in style, and is particularly useful as a rich repository of fairly elementary results.

Whereas ZF is a theory of sets only, both NBG and MK make provision also for classes, i.e. totalities that do not necessarily count as existent mathematical objects. A class-formalism is also provided by Bernays in Reference 69. Bernays' text[69] offers what is in principle a formal axiomatic theory of sets and classes (in the sense of Section 6.3), although the book is written more in the semi-formal manner preferred by working mathematicians.

The theory of ordinals and cardinals has a core which can be developed equally well in ZF, NBG or MK. Sierpiński[73] contains a wealth of classical results, although it is rambling and hard to find things in. Bachmann[74] is more up to date, and good as a reference work.

Most research in axiomatic set theory is concerned, directly or indirectly, with questions of *relative consistency and independence,* i.e. whether certain statements can be added to the basic axioms without creating an inconsistency. On the foundational side, one wants to know how far ZF is internally coherent, and whether certain new axioms could reasonably be adopted. (Gödel's paper 'What is Cantor's continuum problem?' in Benacerraf and Putnam[6] discusses how one might justify adding new axioms to ZF.) On the technical side, one aims to show that conjectures in various branches of mathematics cannot be settled on the basis of some particular set of axioms. (See Section 6.8 for an example.)

The methods used for relative consistency proofs are almost entirely model-theoretic (cf. Section 6.9). One method is to construct an *inner model* (for example, that of the constructible sets) within a given universe of sets. Gödel used this method to prove that the Axiom of Choice and the Continuum Hypothesis could be added to the other axioms of NBG (or ZF) without producing a contradiction; see Krivine[75]. A powerful technique in model theory is *forcing,* which Paul J. Cohen used to show that the negation of the Axiom of Choice can equally well be added to the rest of ZF without producing a contradiction. For the early work on forcing, Felgner[76] is clear and provides good references to the journals; Shoenfield in Reference 11 is smoother but brief. Forcing arguments can usually be recast in terms of *Boolean-valued models* of set theory; Jech[77] makes a good account of the connection between forcing and Boolean-valued models, and gives neat conceptual proofs of many of the hardest and deepest results which have been proved by forcing. More specialised is Jech[78] on the Axiom of Choice.

One axiom which can consistently be added to ZF is Gödel's *axiom of constructibility* ($V = L,$ all sets are constructible). Besides entailing the Continuum Hypothesis, this has various useful combinatorial consequences (cf. Section 6.12). See Mostowski[79] for the older results and Devlin[80] for recent ones. Alternatively, one may consistently add to ZF any one of a family of combinatorial statements known as *Martin's axiom.* These axioms settle many mathematical questions; most of them are inconsistent with $V = L$. See Martin and Solovay[81].

The *large cardinal* axioms also have useful consequences, but there is no hope of showing that they are consistent with ZF. See the full account in Drake[82].

6.8 DESCRIPTIVE SET THEORY

Descriptive set theory is the study of definable sets of real numbers. The founders of the subject in the early years of this century (Borel, Lebesgue, Souslin, Lusin) saw themselves as investigating the structure of the real line—we owe to them the notions of Borel set, Lebesgue measure, Baire category among other things (cf. Section 14.2). But these men also had a constructivist bent: they tended to believe that a set of real numbers must be definable in order to exist. As a result their work is largely a theory of definitions, and inductive definitions in particular. All this early work is assembled in Kuratowski's encyclo-paedic work[83].

For a modern view of inductive definitions, which simultaneously absorbs a large part of descriptive set theory and generalises recursion theory (cf. Section 6.5), see Moschovakis[49]; Barwise[48] is also relevant.

New methods have led to new results on some old problems. One particularly influential example has been *Souslin's problem*. Souslin asked whether there is a complete densely ordered set such that every set of pairwise disjoint open intervals is at most countable, but no countable subset is dense. We now know that there is such a set if $V = L$, but not if Martin's axiom for ω_1 holds; see Devlin and Johnsbråten[84] for a full account. For more new facts on old questions, see Martin in Reference 5.

6.9 MODEL THEORY

Model theory is the theory of mathematical structures and their properties. First-order model theory is concerned with *elementary* properties, i.e. those which can be expressed by means of sets of first-order statements (cf. Section 6.3). For example, the property of being a group is elementary, since the axioms for groups are first-order statements. A set of statements is called a *theory*; a structure is said to be a *model* of a theory if every statement in the theory is true in the model. Model theory has developed a batch of techniques for constructing models of theories, and for ensuring that the models constructed are large or small, thin or fat, rigid or homogeneous, as the occasion demands.

There is an excellent and very thorough textbook of first-order model theory, namely Chang and Keisler[85]. The first half of Bell and Slomson[86] is a highly readable introduction to the subject. Both books handle *ultraproducts*, which are one of the most useful methods of model construction (used by, for example, Ax and Kochen in their partial proof of Artin's conjecture on *p*-adic fields, as described in Chapter 5 of Reference 86). The other most useful method is *indiscernibles* or *Ehrenfeucht–Mostowski models*; this is in Reference 86 but not in Reference 87. The little survey of model theory edited by Morley[87] is good for orientation, and gives further references.

Since Reference 86 was written, there have been two main developments in first-order model theory. The first is the flowering of *stability theory*, mainly in the hands of Shelah[88,89]. This theory seeks to classify theories in terms of the complexity of their models; it leads to some striking results (both positive and negative) on the number of non-isomorphic models of a given size which a theory can have.

The second development is *model-theoretic forcing* (in two forms, finite and infinite) which sprang from the determined efforts of the late Abraham Robinson to capture within model theory the notion of an algebraic closure. The appearance of Robinson's *generic models* coincided providentially with developments in group theory and the theory of skew fields; see Keisler's paper in Reference 88 for a quick summary of the application to group theory, and Hirschfeld and Wheeler[90] for a fuller account of forcing together with the work on skew fields.

Another contribution of Robinson was *non-standard analysis,* which uses model-theoretic methods to simplify analytical and topological arguments. Perhaps the simplest example is Robinson's method for adding infinitesimals to the real line in such a way that the intuitive notion of 'very small' can be used as a precise mathematical concept. See the introduction to the subject by Machover and Hirschfeld[91], or Bernstein's survey in Reference 88, or the collection of essays edited by Luxemburg and Robinson[92]. Robinson's own account[93] is basic but hard to read.

6.10 GENERALISATIONS OF MODEL THEORY

It became clear only within the last 20 years that several classical theorems of first-order model theory remain true when the statements considered are allowed to be infinitely long, provided that some other restrictions are imposed. Thus there arose the model theory of *infinitary languages.* Dickmann[94] is a good general reference, and Keisler[95] is a detailed account of one particular infinitary language. In the early days, interest centred on the connection with large cardinals (cf. Section 6.7) and infinitary combinatorics (cf. Section 6.12). More recently, infinitary languages have become a useful tool for understanding some concepts in algebra; see Barwise in Reference 88.

A few of the more useful theorems of first-order model theory (such as the compactness theorem and the Beth definability theorem) have resisted generalisation to infinitary languages. Logicians have retaliated by isolating axiomatically those languages which do permit good generalisations of these theorems; the result is *abstract model theory*; see Barwise[96] and Makowsky, Shelah and Stavi[96a].

In another direction, Lawvere's success in turning universal algebra (cf. Section 6.11) into category theory has encouraged several people to try to do the same with model theory. The usual setting is an *elementary topos.* Elementary topoi are a generalisation of the presheaf categories of the algebraic geometers of the school of Grothendieck. For Grothendieck topoi, see Artin, Grothendieck and Verdier[97]; for elementary topoi and their links with set theory, see Johnstone[98]; for category-theoretic model theory, see Lawvere, Maurer and Wraith[98a]. Daigneault[99] contains other 'algebraic' views of logic.

6.11   UNIVERSAL ALGEBRA

Universal algebra studies those features which are common to all or most of the classes of structures studied in algebra. For example, many familiar classes of structures are defined by equations; groups are one instance, rings another. A theorem of Birkhoff characterises these classes as those which are closed under isomorphism, substructure, product and homomorphic image—this is a typical theorem of universal algebra. Cohn[100] is a good reference, and Grätzer[101] is a thorough one with a comprehensive bibliography. Jónsson[102] introduces some recent work. See also the recent survey by Taylor[102a].

In so far as it deals with definable classes of structures, universal algebra shows a tendency to be absorbed into model theory: see, for instance, Mal'cev[103]. In so far as it deals with classes defined by equations, universal algebra has largely been swallowed by category theory; see, for example, MacLane[104] *passim,* or Chapters 11 and 18 of Schubert[105]. But there are also parts of the subject which have a life of their own. One such is the theory of lattices; see Grätzer[106]. Even within lattice theory, Boolean algebras are a self-contained topic; see Sikorski[107] and Comfort and Negrepontis[108].

The journal *Algebra Universalis* (Birkhäuser, Basel, 1971-   ) is devoted to universal algebra.

6.12   INFINITARY COMBINATORICS

The field of infinitary combinatorics, concerned with the study of infinite families of sets, is only marginally a branch of logic, although methods from model theory (cf. Section 6.9) have proved useful in recent years. The papers of Kunen and Devlin in Barwise[5] provide a quick summary.

A good deal of work has organised itself around the *partition calculus,* which seeks to generalise Ramsey's theorem on graphs. Most of the known facts are in one or other of Erdös and Rado[109] and Erdös and Hajnal in References 11 and 110. The last two references cover other combinatorial questions as well. Certain large cardinals (cf. Section 6.7) are defined by their partition properties; these include the *weakly compact* and the *Ramsey* cardinals, which have implications in infinitary logic.

Trees, i.e. partial orderings where the predecessors of any element are well-ordered, have provided both problems and methods in combinatorics. Jech[111] is a neat survey from the set-theoretic angle. Trees connect with *ultrafilters*; see Comfort and Negrepontis[108].

Jensen showed that certain useful combinatorial principles hold in the constructible universe; see Devlin[80] for the principles $\Diamond$ and $\Box$. Although these principles may not be true in any absolute sense, they can be used to show that some old conjectures in topology and algebra are not refutable on the basis of accepted set theory.

REFERENCES

1. Shoenfield, Joseph R., *Mathematical logic,* Addison-Wesley, Reading, Mass. (1967)
2. Enderton, Herbert B., *A mathematical introduction to logic,* Academic Press, New York (1972)
3. Mendelson, Elliott, *Introduction to mathematical logic,* Van Nostrand, Princeton (1964)
4. Crossley, J.N. *et al., What is mathematical logic?,* Oxford University Press, Oxford (1972)
5. Barwise, K.J. (ed.), *Handbook of mathematical logic,* North-Holland, Amsterdam (1977)
6. Benacerraf, P. and Putnam, H., *Philosophy of mathematics,* Blackwell, Oxford (1964)
7. van Heijenoort, J., *From Frege to Gödel,* Harvard University Press, Cambridge, Mass. (1967)
8. Hintikka, Jaakko (ed.), *The philosophy of mathematics,* Oxford University Press, Oxford (1969)
9. Kneebone, G.T., *Mathematical logic and the foundations of mathematics,* Van Nostrand, London (1963)
10. Mostowski, Andrzej, *Thirty years of foundational studies,* Blackwell, Oxford (1966)
11. Scott, Dana S. (ed.), *Axiomatic set theory I,* American Mathematical Society, Providence, R.I. (1971)
12. Hilbert, D. and Ackermann, W., *Principles of mathematical logic,* 2nd edn, Chelsea, New York (1950)
13. Kalish, Donald and Montague, Richard, *Logic: techniques of formal reasoning,* Harcourt Brace and World, New York (1964)
14. Lyndon, Roger C., *Notes on logic,* Van Nostrand, Princeton (1966)
15. Beth, Evert W., *Formal methods,* Reidel, Dordrecht (1962)
16. Church, Alonzo, *Introduction to mathematical logic I,* Princeton University Press, Princeton (1956)
17. Gödel, Kurt 'Über formal unentscheidbare Sätze der Principia und verwandter Systeme I' [translated in Davis[39] and van Heijenoort[7]]

18. Hilbert, D. and Bernays, P. *Grundlagen der Mathematik,* 2 vols, 2nd edn, Springer-Verlag, Berlin (1968, 1970)
19. Kleene, Stephen Cole, *Introduction to metamathematics,* Van Nostrand, London (1963)
20. Smullyan, Raymond M., *First-order logic,* Springer-Verlag, Berlin (1968)
21. Prawitz, Dag, *Natural deduction, a proof-theoretic study,* Almqvist and Wiksell, Stockholm (1965)
22. Schütte, Kurt, *Beweistheorie,* Springer-Verlag, Berlin (1960)
23. Kreisel, G., 'A survey of proof theory', *Journal of Symbolic Logic,* **33,** 321-388 (1968)
24. Fenstad, J.E. (ed.), *Proceedings of the Second Scandinavian Logic Symposium,* North-Holland, Amsterdam (1971)
25. Rasiowa, H. and Sikorski, R., *The mathematics of metamathematics,* Państwowe Wydawnictwo Naukowe, Warsaw (1963)
26. Asser, G., 'Theorie der logischen Auswahlfunktionen', *Zeitschrift für Mathematische Logik und Grundlagen der Mathematik,* **3,** 30-68 (1959)
27. Heyting, A., *Intuitionism,* 2nd edn, North-Holland, Amsterdam (1966)
28. van Rootselaar, B. and Staal, J.F. (eds), *Logic, Methodology and Philosophy of Science III,* North-Holland, Amsterdam (1968)
29. Kino, A., Myhill, J. and Vesley, R.E. (eds), *Intuitionism and proof theory,* North-Holland, Amsterdam (1970)
30. Bishop, Errett, *Foundations of constructive analysis,* McGraw-Hill, New York (1967)
31. Lorenzen, Paul, *Differential and integral,* University of Texas Press, Austin (1971)
32. Davis, Martin, *Computability and unsolvability,* McGraw-Hill, New York (1958)
33. Rogers, Hartley, Jr, *Theory of recursive functions and effective computability,* McGraw-Hill, New York (1967)
34. Péter, Rózsa, *Recursive functions,* 3rd edn, Academic Press, New York (1967)
35. Goodstein, R.L., *Recursive number theory,* North-Holland, Amsterdam (1957)
36. Smullyan, Raymond M., *Theory of formal systems,* Annals of Mathematics Studies 47, rev. edn, Princeton University Press, Princeton, N.J. (1961)
37. Hermes, Hans, *Enumerability, decidability, computability,* Springer-Verlag, Berlin (1965)
38. Shepherdson, J.C. and Sturgis, H.E., 'Computability of recursive functions', *Association for Computing Machinery. Journal,* **10,** 217-255 (1963)
39. Davis, Martin (ed.), *The undecidable,* Raven Press, Hewlett, N.Y. (1965)
40. Arbib, Michael A., *Theories of abstract automata,* Prentice-Hall, Englewood Cliffs, N.J. (1969)
41. Ginsburg, Seymour, *The mathematical theory of context-free languages,* McGraw-Hill, New York (1966)
42. Conway, J.H., *Regular algebra and finite machines,* Chapman and Hall, London (1971)
43. Gross, M. and Lentin, A., *Introduction to formal grammars,* Springer-Verlag, Berlin (1970)
44. Manna, Zohar, *Mathematical theory of computation,* McGraw-Hill, New York (1974)
45. Hindley, J.R. Lercher, B. and Seldin, J.P., *Introduction to combinatory logic,* Cambridge University Press, Cambridge (1972)
46. Shoenfield, Joseph R., *Degrees of unsolvability,* North-Holland, Amsterdam (1971)

47. Yates, C.E.M., *A new approach to degree theory* [to appear]
48. Barwise, K.J., *Admissible sets,* Springer-Verlag, Berlin [to appear]
49. Moschovakis, Yiannis N., *Elementary induction on abstract structures,* North-Holland, Amsterdam (1974)
50. Fenstad, J.E. and Hinman, P.G. (eds), *Generalized recursion theory,* North-Holland, Amsterdam (1973)
51. Mazur, S., 'Computable analysis', *Rozprawy Matematyczne,* **33**, 1-111 (1963)
52. Crossley, J.N. and Nerode, Anil, *Combinatorial functors,* Springer-Verlag, Berlin (1974)
53. Davis, Martin, 'Hilbert's tenth problem is unsolvable', *American Mathematical Monthly,* **80**, 233-269 (1973)
54. Suppes, P. *et al.* (eds), *Logic, methodology and philosophy of science IV,* North-Holland, Amsterdam (1973)
55. Rotman, Joseph J., *The theory of groups,* 2nd edn, Allyn and Bacon, Boston (1973)
56. Boone, W.W., Cannonito, F.B. and Lyndon, R.C. (eds), *Word problems,* North-Holland, Amsterdam (1973)
57. Miller, Charles F., III, *On group-theoretic decision problems and their classification,* Annals of Mathematics Studies 68, Princeton University Press, Princeton (1971)
58. Ershov, Yu. L. *et al.,* 'Elementary theories', *Russian Mathematical Surveys,* **20**(4), 35-106 (1965)
59. Tarski, Alfred, Mostowski, Andrzej and Robinson, Raphael M., *Undecidable theories,* North-Holland, Amsterdam (1971)
60. Bar-Hillel, Yehoshua (ed.), *Logic, Methodology and Philosophy of Science, Proceedings of the 1964 International Congress,* North-Holland, Amsterdam (1965)
61. Ackermann, W., *Solvable cases of the decision problem,* North-Holland, Amsterdam (1954)
62. Ax, James, 'The elementary theory of finite fields', *Annals of Mathematics,* **88**, 239-271 (1968)
63. Cohen, Paul J., 'Decision procedures for real and p-adic fields', *Communications on Pure and Applied Mathematics,* **22**, 131-151 (1969)
64. Rabin, M.O., 'Decidability of second-order theories and automata on infinite trees', *American Mathematical Society. Transactions,* **141**, 1-35 (1969)
65. Siefkes, Dirk, *Büchi's monadic second order successor arithmetic,* Lecture Notes in Mathematics 120, Springer-Verlag, Berlin (1970)
66. Reinhardt. W.N., 'Ackermann's set theory equals ZF', *Annals of Mathematical Logic,* **2**, 189-249 (1970)
67. Quine, Willard Van Orman, *Set theory and its logic,* 2nd edn, Harvard University Press, Cambridge, Mass. (1972)
68. Fraenkel, A.A., Bar-Hillel, Y. and Levy, A., *Foundations of set theory,* 2nd edn, North-Holland, Amsterdam (1973)
69. Bernays, P. and Fraenkel, A.A., *Axiomatic set theory,* North-Holland, Amsterdam (1958)
70. Suppes, Patrick, *Axiomatic set theory,* Van Nostrand, Princeton (1960)
71. Monk, J. Donald, *Introduction to set theory,* McGraw-Hill, New York (1969)
72. Takeuti, G. and Zaring, W.M., *Introduction to axiomatic set theory,* Springer-Verlag, New York (1971)

73. Sierpiński, W., *Cardinal and ordinal numbers,* 2nd edn, Polska Akademia Nauk, Warsaw (1965)
74. Bachmann, H., *Transfinite Zahlen,* 2nd edn, Springer-Verlag, Berlin (1967)
75. Krivine, Jean-Louis, *Introduction to axiomatic set theory,* Reidel, Dordrecht (1971)
76. Felgner, Ulrich, *Models of ZF-set theory,* Lecture Notes in Mathematics 223, Springer-Verlag, Berlin (1971)
77. Jech, Thomas J., *Lectures in set theory,* Lecture Notes in Mathematics 217, Springer-Verlag, Berlin (1971)
78. Jech, Thomas J., *The axiom of choice,* North-Holland, Amsterdam (1973)
79. Mostowski, Andrzej, *Constructible sets with applications,* North-Holland, Amsterdam (1971)
80. Devlin, Keith J., *Aspects of constructibility,* Lecture Notes in Mathematics 354, Springer-Verlag, Berlin (1973)
81. Martin, D.A. and Solovay, R.M., 'Internal Cohen extensions', *Annals of Mathematical Logic,* **2,** 143-178 (1970)
82. Drake, Frank R., *Set theory: an introduction to large cardinals,* North-Holland, Amsterdam (1974)
83. Kuratowski, K., *Topology,* 2 vols, Academic Press, New York (1966, 1968)
84. Devlin, Keith J. and Johnsbråten, Håvard, *The Souslin problem,* Lecture Notes in Mathematics 405, Springer-Verlag, Berlin (1974)
85. Chang, C.C. and Keisler, H.J., *Model theory,* North-Holland, Amsterdam (1973)
86. Bell, J.L. and Slomson, A.B., *Models and ultraproducts,* 2nd edn, North-Holland, Amsterdam (1971)
87. Morley, M.D. (ed.), *Studies in model theory,* Mathematical Association of America, Buffalo, N.Y. (1973)
88. Shelah, Saharon, *Stability and the number of non-isomorphic models,* North-Holland, Amsterdam (1977)
89. Shelah, Saharon, 'The lazy model-theoreticians guide to stability', *Logique et Analyse,* 71-72, 241-308 (1975)
90. Hirschfeld, J. and Wheeler, W.H., *Forcing, arithmetic, and division rings,* Lecture Notes in Mathematics 454, Springer-Verlag, Berlin (1975)
91. Machover, M. and Hirschfeld, J., *Lectures on non-standard analysis,* Lecture Notes in Mathematics 94, Springer-Verlag, Berlin (1969)
92. Luxemburg, W.A.J. and Robinson, A. (eds), *Contributions to non-standard analysis,* North-Holland, Amsterdam (1972)
93. Robinson, Abraham, *Non-standard analysis,* North-Holland, Amsterdam (1966)
94. Dickmann, M.A., *Large infinitary languages: model theory,* North-Holland, Amsterdam (1975)
95. Keisler, H. Jerome, *Model theory for infinitary logic: logic with countable conjunctions and finite quantifiers,* North-Holland, Amsterdam (1971)
96. Barwise, K.J., 'Axioms for abstract model theory', *Annals of Mathematical Logic,* 7, 221-226 (1974)
96a. Makowsky, J.A., Shelah, Saharon and Stavi, Jonathan, 'Δ-logics and generalized quantifiers', *Annals of Mathematical Logic,* **10,** 155-192 (1976)
97. Artin, M., Grothendieck, A. and Verdier, J.L., *Théorie des topos et cohomologies étale des schémas,* Lecture Notes in Mathematics 269, Springer-Verlag, Berlin (1972)
98. Johnstone, P.T., *Topos theory,* London Mathematical Society Monographs, Academic Press, London [to appear]
98a. Lawvere, F.W., Maurer, C. and Wraith, G.C., *Model theory and topoi,* Lecture Notes in Mathematics 445, Springer, Berlin (1975)

99. Daigneault, A. (ed.), *Studies in algebraic logic,* Mathematical Association of America, Buffalo, N.Y. (1974)
100. Cohn, P.M., *Universal algebra,* Harper and Row, New York (1965)
101. Grätzer, George, *Universal algebra,* Van Nostrand, Princeton (1968)
102. Jónsson, Bjarni, *Topics in universal algebra,* Lecture Notes in Mathematics 250, Springer-Verlag, Berlin (1972)
102a. Taylor, Walter, 'Equational logic' [to appear]
103. Mal'cev, A.I., *Algebraic systems,* Springer-Verlag, Berlin (1973)
104. MacLane, Saunders, *Categories for the working mathematician,* Springer-Verlag, New York (1971)
105. Schubert, Horst, *Categories,* Springer-Verlag, Berlin (1972)
106. Grätzer, George, *Lattice theory,* Freeman, San Francisco (1971)
107. Sikorski, Roman, *Boolean algebras,* Springer-Verlag, Berlin (1960)
108. Comfort, W.W. and Negrepontis, S., *The theory of ultrafilters,* Springer-Verlag, New York (1974)
109. Erdös, P. and Rado, R., 'A partition calculus in set theory', *American Mathematical Society. Bulletin,* 62, 427-489 (1956)
110. Henkin, Leon (ed.), *Proceedings of the Tarski Symposium,* American Mathematical Society, Providence, R.I. (1974)
111. Jech, Thomas J., 'Trees', *Journal of Symbolic Logic,* 36, 1-14 (1971)

7

Combinatorics

*N.L. Biggs and R.P. Jones**

7.1 INTRODUCTION

It is only in recent years that combinatorial mathematics has attained some cohesion. Among the first practitioners of the subject were famous mathematicians such as Euler (who studied 'partitions') and Cayley (who studied 'trees'); other early workers were well-known mathematical eccentrics, such as Kirkman and Sylvester. The efforts of such men resulted in a subject full of fascinating oddities, but with no order or classification. A good idea of the state of the art at the end of the nineteenth century may be gained from three classic works on mathematical recreations: Lucas[1], Rouse Ball[2], and Ahrens[3].

Early attempts at organisation resulted in the books of Netto[4] and MacMahon[5]; these books are still worth reading, but they give a very one-sided view of the subject as we know it today. In this survey the aim will be to introduce the reader to the main areas of combinatorial mathematics, with special emphasis on those which are centres of current research activity.

There are several journals devoted specifically to this area of mathematics. The *Journal of Combinatorial Theory* (Academic Press, New York, 1966-   ) has been published in two separate series since 1971: Series A contains papers on constructions, designs and

* The authors are indebted to the following people for their helpful comments on a preliminary draft of this chapter: P.J. Cameron, A.W. Ingleton, J.H. van Lint, E.K. Lloyd, R. Rado, R.J. Wilson and D.R. Woodall.

applications, while Series B is devoted mainly to graph theory. *Discrete Mathematics* (North-Holland, Amsterdam, 1971-   ) contains papers on all topics covered in this survey, and some associated subjects. In addition there are two recently founded Journals: *Journal of Graph Theory* (Wiley International, New York, 1977-   ) and *Ars Combinatoria* (University of Waterloo, Waterloo, Ontario, 1975-   ).

Almost all general mathematical journals publish some papers in this field, and these are reviewed in the Combinatorics section of *Mathematical Reviews*. A fairly recent trend is the rapid publication of articles in the less formal manner afforded by Conference Proceedings.

7.2   ENUMERATION

7.2.1   Elementary techniques

Every student of combinatorics should be familiar with such elementary methods as recurrence relations, generating functions and the principle of inclusion and exclusion. There are several good textbooks on general elementary combinatorics; in order of increasing difficulty one can recommend Anderson[6], Ryser[7] and Riordan[8]. These books also contain treatments of several of the famous problems of enumeration, including the problem of 'rencontres' and the 'ménage' problem.

At a more advanced level, a comprehensive reference for basic combinatorial principles is Comtet[9]. In the everyday practice of combinatorics there are two useful works of reference: *Combinatorial identities* by Riordan[10] and *A handbook of integer sequences* by Sloane[11].

7.2.2   Pólya's theory of counting

One of the most powerful techniques in the combinatorialist's armament is the theory of enumeration developed in a long paper by Pólya[12]. This method combines the use of generating functions with ideas from the theory of permutation groups, and it is invaluable in enumerative problems where there are groups of symmetries. Good accounts may be found in the books of Liu[13] and Berge[14]. Pólya's main theorem has been generalised by de Bruijn[15].

The theory of Pólya and de Bruijn is closely interwoven with the classical theory of symmetric functions; see Read[16].

7.2.3   Enumeration of graphs

This branch of enumerative theory is included here since it has few connections with other branches of graph theory (cf. Section 7.4). The subject was begun by Cayley and revitalised by Pólya. A survey of the field is given by Harary and Palmer[17]. Although many kinds of graphs have been counted (for instance, Eulerian graphs by Robinson in Reference 18), there are also many outstanding problems.

A survey of problems associated with the enumeration of labelled trees is given by Moon[19].

7.2.4   Inversion formulae

Let P be a set with a partial ordering $\leqslant$, and suppose that the number of z satisfying $x \leqslant z \leqslant y$ is finite for each x and y in P. Let f and g be functions from P to the reals, such that $f(y)$ and $g(y)$ are zero unless $p \leqslant y$, for some p in P. Then the formula

$$f(y) = \sum_{x \leqslant y} g(x)$$

can be inverted by using the *Möbius function* μ of the partially ordered set, introduced by Rota[20]. The inverted formula takes the form

$$g(y) = \sum_{x \leqslant y} \mu(x, y) f(x)$$

This result contains the principle of inclusion and exclusion, and the well-known Möbius inversion formula of number theory. It is a neat method, but the student should be aware that the same answers can often be obtained by more elementary means.

7.3   COMBINATORIAL SET THEORY

7.3.1   Partition calculus

The simplest and most fundamental result about partitions of a set is Dirichlet's famous pigeon-hole principle: if $n + 1$ objects are distributed among n classes, then at least one class contains more than one object. A sweeping generalisation of this principle was introduced by F.P. Ramsey[21]. Ramsey's results are often formulated in the 'arrow' notation; we write $N \rightarrow (q_1, q_2, \ldots, q_t)^r$ to mean that if A is any set of cardinality N and $A^{(r)}$ is the set of r-element studies of A, then any partition

$$A^{(r)} = K_1 \cup K_2 \cup \ldots \cup K_t$$

satisfies the condition that for some i there is a q_i-element subset X of A such that $X^{(r)}$ is contained in K_i.

For given finite cardinals r, t and $q_1, q_2, \ldots, q_t$, Ramsey's theorem asserts the existence of a finite *Ramsey number* $R = R(r, t, q_1, q_2, \ldots, q_t)$ such that

$$R \to (q_1, q_2, \ldots, q_t)^r$$

For example, $6 \to (3,3)^2$. A good account of many aspects of this theory may be found in a collection of papers by Erdös[22].

For infinite cardinals, Ramsey proved

$$\aleph_0 \to (\aleph_0, \aleph_0, \ldots, \aleph_0)^r$$

A full account of the partition calculus for cardinal numbers is given by Erdös, Hajnal and Rado[23].

7.3.2　Transversals

Let $\mathscr{A} = \{A_i : i \in I\}$ be a finite family of subsets of some fixed finite set. A family $\{x_i : i \in I\}$ is said to be a *transversal* of $\mathscr{A}$ if the x_is are all distinct and x_i belongs to A_i. In 1935 P. Hall gave a simple necessary and sufficient condition (sometimes known as the 'marriage theorem') for the existence of a transversal: the union of any k sets A_i must have at least k members $(1 \leqslant k \leqslant |I|)$. Hall's theorem is a powerful result with applications throughout combinatorics. A good introduction is in Chapter 8 of Wilson[24], and more details may be found in the survey by Mirsky and Perfect[25] and the book by Mirsky[26].

Infinite versions of Hall's theorem have been studied, most recently by Damerell and Milner[27] and Nash-Williams[28].

7.3.3　Intersection theorems

Sperner showed in 1928 that if $\mathscr{F}$ is a family of subsets of an n-element set, with property that no member of $\mathscr{F}$ contains another, then $\mathscr{F}$ has at most

$$\binom{n}{[\tfrac{1}{2}n]}$$

members. There are a great many variants and extensions of this theorem, some of which may be found in the selection of papers by Erdös[22], and an article by Kleitman[29]. For applications to geometry, the student should consult the book by Hadwiger, Debrunner and Klee[30].

7.3.4   Matroids

The study of matroids was initiated by Whitney in 1935. There are several equivalent definitions; one of them is that a *matroid* consists of a set E and a family $\mathscr{B}$ of subsets of E, called *bases*. No base properly contains another base, and if B_1, B_2 are bases and x is in B_1, then there is an element y in B_2 such that $(B_1 - \{x\}) \cup \{y\}$ is a base.

This definition is modelled on the properties of the set of bases of a vector space E. Another example is provided by the set of transversals of a family of subsets of E, and yet another by the family of spanning trees of a connected graph with edge-set E.

A good introduction to matroid theory is the article by Wilson[31]. More advanced treatments are given by Tutte[32] and Welsh[33]; Tutte's book contains the important theory of *graphic* matroids.

In addition to the multitude of possible definitions, there is an alternative terminology, in which a matroid is known as a 'pre-geometry'. This approach is developed at length in a book by Crapo and Rota[34].

7.3.5   Hypergraphs

A *hypergraph* is just a family $\mathscr{E}$ of subsets of a set; when all the members of $\mathscr{E}$ have just two elements, we obtain a graph. In the study of hypergraphs it is customary to use terms derived ultimately from graph theory, despite the fact that hypergraphs have no special structure. The basic text is the book by Berge[35], which is updated in an important survey article by Zykov[36].

One notable success of hypergraph theory was the proof of the 'perfect graph conjecture' by Lovász[31], using techniques from the chromatic theory of hypergraphs; details are given in Berge's book[35].

Several of the topics mentioned in previous sections have been discussed in hypergraph terminology. A survey of such problems is given by Katona[38].

7.4 GRAPH THEORY

7.4.1 Basic graph theory

A *graph* consists of a set of *vertices* and a set of *edges,* together with a relation of *incidence* such that each edge is incident with either one or two vertices. An edge incident with just one vertex is a *loop.* A graph with no loops and with the property that there is at most one edge incident with each pair of vertices is said to be *simple.*

The development of graph theory from 1736 to 1936 is described, with extracts from the classic papers, in the book by Biggs, Lloyd and Wilson[39]; this book may also serve as a text on graph theory. Other, more conventional, texts on basic graph theory are by Behzad and Chartrand[40], Harary[41], and Wilson[24].

There is a considerable amount of published work on elementary properties of paths and circuits in graphs. One perennial problem is to find sufficient conditions for a graph to have a circuit which passes just once through each vertex (a *Hamiltonian* circuit). Material on this problem may be found in the books mentioned above, and in original papers by Pósa[42], Chvátal[43] and Woodall[44].

The famous four-colour conjecture has a vast literature. There are books by Ore[45] and Heesch[46], and a noteworthy recent paper by Tutte and Whitney[47]. In 1976 Appel and Haken[48] announced a proof which makes extensive use of computer calculations.

7.4.2 Colouring problems

The *chromatic number* of a graph is the least number of colours needed so that adjacent vertices may be assigned different colours. The most useful bound for the chromatic number is due to Brooks[49].

G.D. Birkhoff introduced the chromatic polynomial, which gives the number of allowable colourings for each number of colours. Surveys of the properties of this polynomial are given by Read[50] and Biggs[51]. W.T. Tutte[52] has obtained some intriguing results about the zeros of chromatic polynomials of planar graphs.

One may also seek to colour the edges of a graph so that edges with a common vertex are coloured differently. There is an analogue of Brooks's theorem for this situation, due to Vizing[53]. A survey of recent results and conjectures on edge-colourings is given by Wilson in Reference 54.

A edge-colouring of a graph may be regarded as a partition of its edge-set with the property that each part contains no adjacent edges. This approach leads to generalisations of colouring problems, of varying degrees of usefulness. One such generalisation is the problem of *arboricity:* here the parts must contain no circuits. Nash-Williams[55] has given a formula for the arboricity of any graph.

7.4.3 Algebraic graph theory

Both linear algebra and group theory have been used to great effect in the study of graphs.

The *adjacency matrix* of a simple graph with n vertices is the $n \times n$ matrix $A = (a_{ij})$ in which $a_{ij} = 1$ if the vertices i and j are adjacent, and $a_{ij} = 0$ otherwise. The eigenvalues of A yield a certain amount of useful information about the graph, but they do not determine it. An introduction to this field may be found in the book by Biggs[51].

An *automorphism* group of a graph is a permutation of its vertex-set which preserves adjacency. The set of automorphisms is a group with the operation of composition. In 1938 Frucht proved that every abstract group is the group of some graph, and this result remains true if we stipulate that the graphs must have some assigned properties—see Izbicki[56]. It is more instructive to study the group of automorphisms as a permutation group, rather than as an abstract group.

We may require the automorphism to act transitively on the vertices, or on pairs of adjacent vertices, or on pairs of vertices at each given distance. This hierarchy of symmetry conditions gives rise to much interesting mathematics—see Biggs[51]. In the last of the three cases mentioned, the graph has the property known as *distance-regularity*; special cases include the *Moore graphs* (Hoffman and Singleton[57], Damerell[58]) and the *strongly regular* graphs surveyed by Hubaut[59].

7.4.4 Extremal problems

One aspect of extremal theory arises from the applications of Ramsey's theorem (cf. subsection 7.3.1) to graphs. For example, taking $r = t = 2$ and $q_1 = q_2 = q$, we deduce that in any 2-colouring of the edges of the complete graph with $R(2,2,q,q)$ vertices there will be a monochromatic complete subgraph with q vertices. Erdös[22] has written many papers on such questions; see also Graver and Yackel[60]. Harary (in Reference 61) and others have considered the Ramsey number $R(G_1,G_2)$ of two graphs. This is defined to be the smallest number n such that in any 2-colouring of the edges of the complete graph on n vertices there will be a G_1 with one colour or a G_2 with the other.

Another type of extremal problem is exemplified by Turán's theorem[62]: a graph with n vertices and at least $[n^2/4]$ edges must contain a triangle. A survey of such problems is given by Erdös in Reference 63.

Closely allied to extremal theory is the use of probabilistic methods in graph theory. For example, probabilistic reasoning may be used to show that there are graphs whose girth and chromatic number are arbitrarily large. The student should consult the book by Erdös and Spencer[64] for details.

7.4.5 Directed graphs

In a *directed graph* (or *digraph*) the vertices are joined by *arcs* having an assigned direction; in other words, an arc is an element of $V \times V$, where V is the vertex-set. The basic theory of digraphs is very similar to graph theory, and accounts of it may be found in the textbooks mentioned in subsection 7.4.1.

A digraph whose underlying graph is a complete graph is sometimes called a *tournament*. There is a stimulating book by Moon[65] on this subject.

7.4.6 Topological graph theory

A graph may be represented by a topological space in which the vertices are represented by points and the edges are represented by lines joining the points. Topological graph theory is concerned with the problem of imbedding the representative space in certain surfaces, in such a way that the lines intersect only at their assigned end points. For example, a graph is said to be *planar* if it can be so imbedded in the plane or sphere.

Kuratowski[66] gave a combinatorial classification of non-planar graphs, but there are few similar results for other surfaces. For a graph imbedded in an orientable surface of genus g the number of faces is given by the *Euler characteristic* formula '$V - E + F = 2 - 2g$'. The *genus* $\gamma(G)$ of a graph G is defined to be the smallest g such that G can be imbedded in the surface of genus g, and the formula yields a lower bound for $\gamma(G)$. For the complete graph K_n we get

$$\gamma(K_n) \geqslant \{\tfrac{1}{12}(n-3)(n-4)\}$$

The Heawood conjecture was that equality held in the above; this was finally proved in 1967—see Ringel[67] for a full account. An interesting account of some algebraic aspects of the theory of graph imbeddings may be found in the book by White[68].

7.4.7   Miscellaneous topics

A *matching* of a graph is a subset of its edge-set with the property that no two edges have a common vertex. The matching is *perfect* (or a *1-factor*) if it covers every vertex. Tutte[69] found a necessary and sufficient condition for a connected graph to have a 1-factor; a very short proof using Hall's theorem is given by Anderson[70]. An alternative approach to matching problems uses the so-called 'Hungarian' method of alternating chains—see Berge[35] for details.

In 1927 Menger proved an important result concerning the 'connectivity' of a graph. His result implies Hall's theorem and it is frequently employed in transversal theory; see Wilson[24] for an elementary account.

7.5   FINITE GEOMETRIES AND DESIGNS

7.5.1   General theory

A $t-(v,k,\lambda)$ *design* is a collection B of subsets of a set S, with $|S| = v$, satisfying two conditions: (1) each member of B has cardinality k; (2) each set of t elements of S is contained in exactly λ members of B. The sets in B are called *blocks* and the elements of S are called *points*. The possibility of 'repeated blocks' is frequently not allowed; that is, all members of the collection B are different subsets of S. When $\lambda = 1$, the design is said to be a *Steiner system*. In order to exclude degenerate and uninteresting cases it is usually assumed that $v \geqslant k \geqslant t \geqslant 2$ and $\lambda \neq 0$.

The study of designs originated in the statistical theory of experiments, during the 1930s, but it was not until the important paper of Bruck and Ryser[71] in 1949 that mathematicians became interested in the subject. Nowadays design theory has infiltrated many areas of classical geometry—especially the study of geometry in projective and vector spaces over finite fields. General references are the books by Hall[72] and Dembowski[73]. An important source of fundamental ideas is the work of Tits[74].

7.5.2   Designs with $t \geqslant 3$

A *t-design* is a $t-(v,k,\lambda)$ design for some parameters (v,k,λ). For $t \geqslant 3$ there is very little general theory. Some examples are known, and since these tend to be linked with unexplained phenomena in other subjects—such as group theory—there is considerable interest in them. Some generally applicable criteria for the existence of such designs are the so-called divisibility criteria (see Dembowski[73]) but these necessary conditions are by no means sufficient.

For $t > 5$ the only known designs without repeated blocks are the designs in which every set of k points is a block. For $t = 4$ and $t = 5$ there were only four known designs until quite recently: they are Steiner systems with parameters $4 - (11,5,1)$, $5 - (12,6,1)$, $4 - (23,7,1)$ and $5 - (24,8,1)$, discussed by Witt[75]. Some new 5-designs with $\lambda \neq 1$ have been discovered in a coding theory context, and a good account of them is given by Cameron and van Lint[76]. Very recently some new 5-designs with $\lambda = 1$ have been found by R.H.F. Denniston[77]. The parameters of two of Denniston's designs are $5 - (28,7,1)$ and $5 - (48,6,1)$.

For $t = 3$ infinitely many designs are known to exist. Examples are the *Möbius planes* with parameters $3 - (q^2 + 1, q + 1, 1)$, where q is a prime power (see Dembowski[73]).

7.5.3 Designs with $t = 2$

A 2-design in which not every k-subset of S is a block is a classical *balanced incomplete block design* (BIBD). A comprehensive treatment of such designs is given in the book by Hall[72]. This book also gives useful tables of 2-designs for small values of the parameters, but the reader should be aware that several designs listed as 'not known' have since been discovered; Hall in Reference 78 gives more up-to-date information.

The basic tool in the study of 2-designs is the incidence matrix. For example, if b denotes the number of blocks, then the inequality $b \geq v$ can be proved by simple matrix arguments. When $b = v$, the design is said to be *symmetric*. There are important necessary conditions for the existence of a symmetric $2 - (v,k,\lambda)$ design, known as the Bruck–Ryser–Chowla theorem: if v is even, then $k - \lambda$ must be a square, while if v is odd, the equation

$$x^2 = (k-\lambda)y^2 + (-1)^{\frac{1}{2}(v-1)}\lambda z^2$$

must have a solution in integers x,y,z, not all zero. A straightforward account is given by Biggs[79].

7.5.4 Projective planes

A particularly interesting class of 2-designs are the *projective planes*, which have parameters $2 - (n^2 + n + 1, n + 1, 1)$. The books of Dembowski[73] and Hall[72] mentioned previously contain a great deal of material on this topic; a more recent book is by Hughes and Piper[80].

The number n occurring in the definition is said to be the *order* of the plane. For each prime power order q there is a standard example, constructed in the classic manner as a 2-dimensional projective space over the finite field $GF(q)$. Other examples are known, but they all have prime power order. The possibility of a projective plane of order 10 is a matter of much current speculation—see MacWilliams, Sloane and Thompson[81] for an approach through coding theory.

There has been much work on the structure of the group of collineations of projective planes. A complete classification of the possibilities is given by Dembowski[73] and is known as the Lenz–Barlotti scheme. Almost all known planes are of the kind known as 'translation planes'; these are the subject of lecture notes by Ostrom[82].

7.5.5   Latin squares

Latin squares are another aspect of combinatorial theory which finds application in the design of experiments. There is a mass of heterogeneous information on diverse aspects of the subject, including the relationship with projective planes, and the reader is fortunate in being able to refer to the treatise by Dénes and Keedwell[83] for a comprehensive account.

7.5.6   Other topics

There are several other topics, variously linked with design theory, of which the student should be aware. Among these are Hadamard matrices and Room squares (Wallis, Street and Wallis[84]), generalised polygons (Dembowski[73]) and biplanes (Cameron[85]).

7.6   CODING THEORY

7.6.1   Theory and applications

Practical coding techniques depend on two mathematical theories— information theory and combinatorial theory. The latter has led to the formulation of several new concepts, and to the establishment of links with designs and graphs.

In the beginning (about 1950) the setting for combinatorial coding theory was a vector space F^n of dimension n over the field $F = GF(2)$. A subspace C of F^n is a *linear binary e-code* (*e*-error correcting code) if any two vectors in C differ in at least $2e + 1$ coordinates; it follows that any vector in F^n differs in e (or fewer) coordinates from at most one vector in C.

The first book on coding theory was written by Peterson[86] in 1961. There are more recent books by Berlekamp[87], which concentrates on cyclic codes, and by Blake and Mullin[88], which contains rather too much unnecessary material. Each of these books gives an account of the numerous methods for constructing binary codes with given properties. For a survey of results concerning the 'best' codes, in certain practical senses, the student should consult Sloane[89].

Several generalisations of the classical model are possible, and these lead to questions of considerable mathematical interest but little practical application. Such questions will be considered in the following sections.

7.6.2 Perfect codes

The field $GF(2)$ may be replaced by the general finite field $GF(q)$, q a prime power, and the subset C need not be a subspace. In this case, C is said to be a *perfect e*-code if any vector in F^n differs in e (or fewer) coordinates from exactly one vector in C. A very readable introduction to the mathematical aspects of this situation is given by van Lint[90]. It is remarkable that there are very few perfect codes; apart from some trivial cases and some families of perfect 1-codes (called the Hamming codes) there are just two perfect codes: (1) a linear space of dimension 11 in F^{23}, $F = GF(2)$; (2) a linear space of dimension 6 in F^{11}, $F = GF(3)$. This result is due to Tietäväinen[91]. Tietäväinen's proof relies on a computer check to eliminate some 'small' cases, and the need for this has recently been removed by Smith[92]. Applications of the classification of binary perfect codes may be found in the work on parallelisms by Cameron[93].

7.6.3 Other developments

The perfect code question has led to the formulation of new, more general, settings for coding theory. Delsarte[94] introduced the theory of association schemes and used methods from linear programming; Biggs[95] studied perfect codes in graphs. In both contexts it is possible to prove a general version of the theorem of Lloyd (see van Lint[90]), which gives a necessary condition for the existence of a perfect code in terms of the zeros of an associated polynomial.

Slight generalisations of the notion of a perfect code—'nearly perfect' and 'uniformly packed' codes—are useful in studying other combinatorial problems. The best reference for such matters is the book of Cameron and van Lint[76].

Another important development is the use of invariant theory to study the weight enumerators of codes; this may be found in papers by Gleason[96] and Sloane[97].

7.7   APPLICATIONS

There are very many diverse applications of combinatorial theory, so that it is impossible to describe all of them. Some of the more important fields of application have led to developments in the central theory; four such fields will be mentioned here.

7.7.1   Electrical network theory

The study of electrical networks led Kirchhoff to discover some basic theorems of graph theory as long ago as 1847. There are now several books describing the relevant electrical engineering techniques in terms of matrix theory—for example, Seshu and Reed[98] and Chen[99]. The student should not attempt a detailed study of these books, but he should be aware of their contents.

One noteworthy development is the use of the 'topological formulae' due to Mason and Coates for the solution of network equations. Chen's book contains an extended account of this.

7.7.2   Flows in networks

When the variable whose flow is being measured is constrained by a 'capacity' function, the theory is rather different from the flow of current in an electrical network. The basic reference is the book by Ford and Fulkerson[100]. This book contains an excellent account of the theoretical connections with Hall's theorem and the practical applications to problems such as assignment and transportation.

7.7.3   Permutation groups

The study of an abstract group, and sometimes its very definition, may be facilitated by regarding the group as operating on some assigned structure. The automorphisms of finite geometries, designs, codes and graphs have all been investigated for this reason. An introduction to this field is given by Biggs[79]; see also Kantor[101].

A great deal of interest centres on the construction of finite simple groups. Higman and Sims[102] discovered a new simple group as a group of automorphisms of a graph with 100 vertices, and since that time over a dozen 'sporadic' simple groups have been constructed in a similar fashion. A survey is given by Tits in Reference 103.

7.7.4 Models of physical phenomena

The prototype for combinatorial models of physical phenomena is the famous Ising model of ferromagnetism. Such models are concerned, roughly speaking, with the behaviour of functions whose domain of definition is a set of 'states' on a graph. In the 'thermodynamic limit', as the size of the graph increases, these functions may exhibit singularities, which indicate that a 'phase transition' occurs. A survey of the whole field, including many stimulating applications of combinatorial methods, may be found in a series of volumes edited by Domb and Green[104]. The book by Thompson[105] is also recommended.

REFERENCES

1. Lucas, E., *Récréations mathématiques,* Gauthier-Villars, Paris (1882)
2. Rouse Ball, W.W., *Mathematical recreations and problems of past and present times,* Macmillan, London (1892)
3. Ahrens, W., *Mathematische Unterhaltungen und Spiele,* Teubner, Leipzig (1901)
4. Netto, E., *Lehrbuch der Combinatorik,* Teubner, Leipzig (1901)
5. MacMahon, P.A., *Combinatory analysis,* Cambridge University Press, Cambridge (1916)
6. Anderson, I., *A first course in combinatorial mathematics,* Oxford University Press, Oxford (1974)
7. Ryser, H.J., *Combinatorial mathematics,* Carus Mathematical Monographs 14, Mathematical Association of America, Buffalo, N.Y. (1963)
8. Riordan, J., *An introduction to combinatorial analysis,* Wiley, New York (1958)
9. Comtet, L., *Advanced combinatorics: the art of finite and infinite expansions,* rev. edn. Reidel, Dordrecht (1974)
10. Riordan, J., *Combinatorial identities,* Wiley, New York (1968)
11. Sloane, N.J.A., *A handbook of integer sequences,* Academic Press, New York (1973)
12. Pólya, G., 'Kombinatorische Anzahlbestimmungen für Gruppen, Graphen und chemische Verbindungen', *Acta Mathematica,* 68, 145-254 (1937)
13. Liu, C.L., *Introduction to combinatorial mathematics,* McGraw-Hill, New York (1968)
14. Berge, C., *Principles of combinatorics,* Academic Press, New York (1971)

15. de Bruijn, N.G., 'Generalisations of Pólya's fundamental theorem in enumerative combinatory analysis', *Indagationes Mathematicae*, **21**, 59-69 (1959)
16. Read, R.C., 'The use of S-functions in combinatorial analysis', *Canadian Journal of Mathematics*, **20**, 808-841 (1968)
17. Harary, F. and Palmer, E.M., *Graphical enumeration*, Academic Press, New York (1973)
18. Harary, F. (ed.), *Proof techniques in graph theory*, Academic Press, New York (1969)
19. Moon, J.W., *Counting labelled trees*, Canadian Mathematical Monographs 1, Canadian Mathematical Congress, Montreal (1970)
20. Rota, G.C., 'On the foundations of combinatorial theory I: Theory of Möbius functions', *Zeitschrift für Wahrscheinlichkeitstheorie und Verwandte Gebiete*, **2**, 340-368 (1964)
21. Ramsey, F.P., 'On a problem of formal logic', *London Mathematical Society. Proceedings*, **30**, 264-286 (1930)
22. Erdös, P., *The art of counting—selected writings*, M.I.T. Press, Cambridge, Mass. (1973)
23. Erdös, P., Hajnal, A. and Rado, R., 'Partition relations for cardinal numbers', *Acta Mathematica Academiae Scientiarium Hungaricae*, **15**, 93-196 (1965)
24. Wilson, R.J., *Introduction to graph theory*, Oliver and Boyd, Edinburgh (1972)
25. Mirsky, L. and Perfect, H., 'Systems of representatives', *Journal of Mathematical Analysis and Applications*, **15**, 520-568 (1966)
26. Mirsky, L., *Transversal theory*, Academic Press, New York (1971)
27. Damerell, R.M. and Milner, E.C., 'Necessary and sufficient conditions for transversals of countable set systems', *Journal of Combinatorial Theory. A*, **17**, 350-374 (1974)
28. Nash-Williams, C. St. J.A., 'Marriage in denumerable societies', *Journal of Combinatorial Theory. A*, **19**, 335-336 (1975)
29. Kleitman, D.J., 'On an extremal property of antichains in partial orders, the LYM property and some of its implications and applications', *Mathematical Centre Tracts*, **56**, 77-90 (1974)
30. Hadwiger, H., Debrunner, H. and Klee, V., *Combinatorial geometry in the plane*, Holt, Rinehart and Winston, New York (1964)
31. Wilson, R.J., 'An introduction to matroid theory', *American Mathematical Monthly*, **80**, 500-525 (1973)
32. Tutte, W.T., *Introduction to the theory of matroids*, American Elsevier, New York (1971)
33. Welsh, D.J.A., *Matroid theory*, Academic Press, London (1976)
34. Crapo, H.H. and Rota, G.C., *On the foundations of combinatorial theory: combinatorial geometries*, M.I.T. Press, Cambridge, Mass. (1970)
35. Berge, C., *Graphs and hypergraphs*, North-Holland, Amsterdam (1973)
36. Zykov, A.A., 'Hypergraphs', *Russian Mathematical Surveys*, **29**, 89-156 (1974)
37. Lovász, L., 'Normal hypergraphs and the perfect graph conjecture', *Discrete Mathematics*, **2**, 253-267 (1972)
38. Katona, G.O.H., 'Extremal problems for hypergraphs', *Mathematical Centre Tracts*, **56**, 13-42 (1974)
39. Biggs, N.L., Lloyd, E.K. and Wilson, R.J., *Graph theory 1736-1936*, Oxford University Press, Oxford (1976)
40. Behzad, M. and Chartrand, G., *Introduction to the theory of graphs*, Allyn and Bacon, Boston (1971)

41. Harary, F., *Graph theory,* Addison-Wesley, Reading, Mass. (1969)
42. Pósa, L., 'A theorem concerning Hamilton lines', *Magyar Tudományos Akadémia: Matematikai Kutató Intézet. Közleményei,* 7, 225-226 (1962)
43. Chvátal, V., 'On Hamilton's ideals', *Journal of Combinatorial Theory. B,* 12, 163-168 (1972)
44. Woodall, D.R., 'Sufficient conditions for circuits in graphs', *London Mathematical Society. Proceedings',* 24, 739-755 (1975)
45. Ore, O., *The four-color problem,* Academic Press, New York (1967)
46. Heesch, H., *Untersuchungen zum Vierfarbenproblem,* Bibliographisches Institut, Mannheim (1969)
47. Tutte, W.T. and Whitney, H., 'Kempe chains and the four colour problem', *Utilitas Mathematica,* 2, 241-281 (1972)
48. Appel, K. and Haken, W., 'Every planar map is four-colorable, *American Mathematical Society. Bulletin,* 82, 711-712 (1976)
49. Brooks, R.L., 'On colouring the nodes of a network', *Cambridge Philosophical Society. Proceedings,* 37, 194-197 (1941)
50. Read, R.C., 'An introduction to chromatic polynomials', *Journal of Combinatorial Theory,* 4, 52-71 (1968)
51. Biggs, N.L., *Algebraic graph theory,* Cambridge University Press, Cambridge (1974)
52. Tutte, W.T., 'On chromatic polynomials and the golden ratio', *Journal of Combinatorial Theory,* 9, 289-296 (1970)
53. Vizing, V.G., 'On an estimate of the chromatic class of a p-graph', *Diskretnyi Analiz,* 3, 25-30 (1964)
54. Alavi, Y. and Lick, D.R. (eds.), *Theory and Applications of Graphs,* Lecture Notes in Mathematics, Springer-Verlag, Berlin (1977)
55. Nash-Williams, C. St. J.A., 'Edge-disjoint spanning trees of finite graphs', *London Mathematical Society. Journal,* 36, 445-450 (1961)
56. Izbicki, H., 'Unendliche Graphen endlichen Grades mit vorgegebenen Eigenschaften', *Monatshefte für Mathematik,* 63, 298-301 (1960)
57. Hoffman, A.J. and Singleton, R.R., 'On Moore graphs with diameters 2 and 3', *IBM Journal of Research and Development,* 4, 497-504 (1960)
58. Damerell, R.M., 'On Moore graphs', *Cambridge Philosophical Society, Proceedings,* 74, 227-236 (1973)
59. Hubaut, X.L., 'Strongly regular graphs', *Discrete Mathematics,* 13, 357-382 (1975)
60. Graver, J.E. and Yackel, J., 'Some graph theoretic results associated with Ramsey's Theorem', *Journal of Combinatorial Theory,* 4, 125-175 (1968)
61. Nash-Williams, C. St. J.A. and Sheehan, J. (eds), *Proceedings of the Fifth British Combinatorial Conference, 1975,* Utilitas Mathematica, Winnipeg
62. Turán, P., 'Eine Extremalaufgabe aus der Graphentheorie', *Matematikai és Fizikai Lapok,* 48, 436-452 (1941)
63. Harary, F. (ed.), *A seminar in graph theory,* Holt, Rinehart and Winston, New York (1967)
64. Erdös, P. and Spencer, J., *Probabilistic methods in combinatorics,* Academic Press, New York (1974)
65. Moon, J.W., *Topics on tournaments,* Holt, Rinehart and Winston, New York (1968)
66. Kuratowski, K., 'Sur le problème des courbes gauches en topologie', *Fundamenta Mathematicae,* 15, 271-283 (1930)
67. Ringle, G., *Map color theorem,* Springer-Verlag, Berlin (1974)
68. White, A.T., *Graphs, groups and surfaces,* North-Holland, Amsterdam (1973)

69. Tutte, W.T., 'The factorization of linear graphs', *London Mathematical Society. Journal,* **22**, 107-111 (1947)
70. Anderson, I., 'Perfect matchings of a graph', *Journal of Combinatorial Theory,* **10**, 183-186 (1971)
71. Bruck, R.H. and Ryser, H.J., 'The nonexistence of certain finite projective planes', *Canadian Journal of Mathematics,* **1**, 88-93 (1949)
72. Hall, M., *Combinatorial theory,* Blaisdell, Waltham, Mass. (1967)
73. Dembowski, P., *Finite geometries,* Springer-Verlag, Berlin (1968)
74. Tits, J., *Buildings of spherical type and finite BN-pairs,* Lecture Notes in Mathematics 386, Springer-Verlag, Berlin (1974)
75. Witt, E., 'Über Steinersche System', *Hamburg, Universität: Mathematische Seminar. Abhandlungen,* **12**, 265-275 (1938)
76. Cameron, P.J. and van Lint, J.H., *Graph theory, coding theory, and block designs,* London Mathematical Society Lecture Note Series 19, Cambridge University Press, Cambridge (1975)
77. Denniston, R.H.F., 'Some new 5-designs', *London Mathematical Society. Bulletin,* **8**, 263-267 (1976)
78. Srivastava, J.N. (ed.), *A survey of combinatorial theory,* North-Holland, Amsterdam (1973)
79. Biggs, N.L., *Finite groups of automorphisms,* London Mathematical Society Lecture Note Series 6, Cambridge University Press, Cambridge (1971)
80. Hughes, D.R. and Piper, F.C., *Projective planes,* Springer-Verlag, New York (1973)
81. MacWilliams, F.J., Sloane, N.J.A. and Thompson, J.G., 'On the existence of a projective plane of order 10', *Journal of Combinatorial Theory. A,* **14**, 66-78 (1973)
82. Ostrom, T.G., *Finite translation planes,* Lecture Notes in Mathematics 158, Springer-Verlag, Berlin (1970)
83. Dénes, J. and Keedwell, A.D., *Latin squares and their applications,* Akadémiai Kiadó, Budapest (1974)
84. Wallis, W.D., Street, A.P. and Wallis, J.S., *Combinatorics: Room squares, sum-free sets, Hadamard matrices,* Lecture Notes in Mathematics 292, Springer-Verlag, Berlin (1972)
85. Cameron, P.J., 'Characterisations of some Steiner systems, parallelisms and biplanes', *Mathematische Zeitschrift,* **136**, 31-39 (1974)
86. Peterson, W.W., *Error correcting codes,* M.I.T. Press, Cambridge, Mass. (1961)
87. Berlekamp, E.R., *Algebraic coding theory,* McGraw-Hill, New York (1968)
88. Blake, I. and Mullin, R.C., *The mathematical theory of coding,* Academic Press, New York (1975)
89. Sloane, N.J.A., 'A survey of constructive coding theory, and a table of binary codes of highest known rate', *Discrete Mathematics,* **3**, 265-294 (1972)
90. Lint, J.H. van, *Coding theory,* Lecture Notes in Mathematics 201, Springer-Verlag, Berlin (1971)
91. Tietäväinen, A., 'On the non-existence of perfect codes over finite fields', *SIAM Journal on Applied Mathematics,* **24**, 88-96 (1973)
92. Smith, D.H., 'An improved version of Lloyd's theorem', *Discrete Mathematics,* **15**, 175-184 (1976)
93. Cameron, P.J., *Parallelisms of complete designs,* London Mathematical Society Lecture Note Series 23, Cambridge University Press, Cambridge (1976)
94. Delsarte, P., 'The association schemes of coding theory', *Mathematical Centre Tracts,* **55**, 139-157 (1974)

95. Biggs, N.L., 'Perfect codes in graphs', *Journal of Combinatorial Theory. B*, **15**, 289-296 (1973)

96. Gleason, A.M., 'Weight polynomials of self-dual codes and the MacWilliams identities', *International Congress of Mathematicians. Proceedings*, **3**, 211-215 (1970)

97. Sloane, N.J.A., 'Weight enumerators of codes', *Mathematical Centre Tracts*, **55**, 111-138 (1974)

98. Seshu, S. and Read, M.B., *Linear graphs and electrical networks*, Addison-Wesley, Reading, Mass. (1961)

99. Chen, W.-K., *Applied graph theory*, North-Holland, Amsterdam (1971)

100. Ford, L.R. and Fulkerson, D.R., *Flows in networks*, Princeton University Press, Princeton (1962)

101. Kantor, W.M., '2-transitive designs', *Mathematical Centre Tracts*, **57**, 44-97 (1974)

102. Higman, D.G. and Sims, C.C., 'A simple group of order 44,352,000', *Mathematische Zeitschrift*, **105**, 110-113 (1968)

103. *Séminaire Bourbaki. Exposés 364-381*, Lecture Notes in Mathematics 180, Springer-Verlag, Berlin (1971)

104. Domb, C. and Green, M.S., *Phase transitions and critical phenomena*, 6 vols, Academic Press, New York (1972-1976)

105. Thompson, C.J., *Mathematical statistical mechanics*, Macmillan, New York (1972)

8

Rings and Algebras

P.M. Cohn

8.1   INTRODUCTION

Abstraction is the hall-mark of modern algebra. If we examine the operations and laws of the usual number system, but abstract from the nature of numbers, we reach the notion of a *ring*, or a *field* if division is permitted. The development of ring theory has been in two directions: On the one hand, such instances as the quaternions and matrix algebras gave rise to the subject of linear associative (and sometimes non-associative) algebras, which developed into the study of Artinian and more recently Noetherian rings. On the other hand, there are the rings of algebraic integers; they are essentially Dedekind rings, which arose also in algebraic geometry as coordinate rings of curves and which formed the beginning of a very active field of study—commutative ring theory, which especially in the last few decades has had a powerful impetus from algebraic geometry.

We shall pass over an even more basic notion, that of a group, since it has a separate chapter (cf. Chapter 9). But several other notions are deserving of mention. Various types of non-associative algebras arise quite naturally in different contexts, foremost among them *Lie algebras* with close links to groups (both infinite and finite). In studying the structure underlying a group or algebra one is led to the notion of a lattice, which in turn has close links to Boolean algebras in logic and to ordered algebraic systems (cf. Section 6.11). In a more profound analysis of groups and algebras one abstracts from the notion of group and mapping. This leads to the concept of a *category*, which provides a convenient language for much of algebra and indeed of other parts of mathematics, besides being a subject of study in its own right. It arose from the methods of homological algebra first developed in the 1940s to describe techniques in algebraic topology, but which since then have become quite indispensable for algebra itself.

It goes without saying that complete coverage in a brief article of this kind is out of the question; the selection made inevitably reflects the predilections (and limited knowledge) of the author, but some attempt has been made to mention most major areas of research.

8.2 ASSOCIATIVE RINGS

Most undergraduate courses include some discussion of rings—in particular, the example of the integers Z and the ring $k[x]$ of polynomials in an indeterminate x over a field k. Among the properties shared by Z and $k[x]$ the *Euclidean algorithm* plays a prominent role; apart from its intrinsic interest, it serves to establish the uniqueness of factorisation and the fact that these rings are principal ideal domains. Many basic properties of rings can be illustrated by these examples, e.g. congruences lead to the notion of residue class ring; at a more advanced level the completion of a local ring is illustrated by the example of the p-adic numbers. The more formal aspects of vector space theory can be stated in the setting of modules over a general ring. An invariant (i.e. coordinate-free) treatment of vector spaces leaves one better prepared to work with modules, where no basis need exist. A first course will deal with general properties, e.g. the isomorphism theorems, which may be stated for groups with operators, so as to apply to modules and, when suitably modified, to rings themselves. This may be followed by the classification of modules over a principal ideal domain, itself useful for the similarity reduction of matrices. All these matters are treated in most undergraduate texts, e.g. Jacobson[1], Herstein[2], Lang[3] and Cohn[4], and are tacitly assumed known in any more advanced discussion.

Matrices, especially over the real and complex numbers, have been studied very thoroughly, but are not the concern of this chapter. However, a brief mention must be made of the recent work by Gelfand, Ponomarev and others on treating classification problems of linear algebra by diagrams analogous to the Coxeter–Dynkin diagrams for semisimple Lie algebras. See Bernstein, Gelfand and Ponomarev[5,6], Gabriel[7,8], Dlab and Ringel[9] and Ringel[10].

A central part of ring theory is the classification of semisimple rings—in particular, the Wedderburn theorems. There are two main approaches. The first is via semisimple (i.e., completely reducible) modules, possibly using homological algebra. The other is via the

Jacobson structure theory, which allows the result to be derived with a minimum of assumptions. The Jacobson radical can be defined for any ring (even without a 1) and for Artinian rings this reduces to the set of properly nilpotent elements. The density theorem applies to any primitive ring, but for Artinian rings again yields Wedderburn's theorems. One of the main applications is to the representation theory of finite groups in characteristic 0 (or at least prime to the group order). For a thorough account see Curtis and Reiner[11], but the basic notions can be found in most graduate texts, e.g. Lang[3] or Cohn[12].

An important development is the theory of Artinian rings which are not semisimple. Such rings arise in the study of group representations over fields of characteristic dividing the group order (modular representations). Here the group algebra has a duality which has been studied abstractly under the name of *Frobenius algebra*. A weaker form, the *quasi-Frobenius ring* (an Artinian ring in which each one-sided ideal is an annihilator) has been the subject of independent study; see Curtis and Reiner[11]. Much work has been done in this more general context, especially the study of indecomposable modules; see Roiter[13].

Just as the integers served as the prototype of a ring, the rational numbers form the model of a field. Although algebraic number fields were much studied in the nineteenth century, the foundations for a general study were laid in a classic paper by Steinitz[14]. Much of this, together with Galois theory, is in most current textbooks (e.g. Jacobson[15], Lang[3] or Cohn[12]). Further developments in field theory are in the direction of function theory (Chevalley[16]), algebraic geometry (Šafarevič[17]) or number theory (Lang[18]). There is also much older work waiting to be taken up again, e.g. on equations with icosahedral group; an excellent source for this material is Weber[19]. In some of these questions, e.g. Noether's problem (determination of the fixed field of a permutation group acting on a purely transcendental extension), there has been considerable progress lately; for a very lucid survey by one of the main contributors see Lenstra[20]. Galois theory itself has been generalised to deal with certain extensions of commutative rings, by Chase, Harrison and Rosenberg[21].

Much less is known about skew fields (i.e. not necessarily commutative division rings), although they arise quite naturally in Schur's lemma and the density theorem. Most of what is known refers to skew fields finite-dimensional over their centre. Such division algebras are a special case of central simple algebras which were studied intensively in the 1930s. Central simple algebras over a given field k are divided into classes of similar algebras, and these classes form a group $\mathbf{B}_k$, the *Brauer group* of k, which turns out to be a basic homological invariant of k. An excellent though condensed account is in Deuring[22]; for a more modern treatment see Serre[23], Weil[24] or Reiner[25]. The structure of $\mathbf{B}_Q$ can be completely determined, but this requires some results from class-field theory (cf. Cassels and Fröhlich[26]). Most accounts are content to assume these results, an exception being Weil[24], which is self-contained. Using homological methods, one can define the Brauer group for any commutative ring A (Auslander and Goldman[27]) as a group of similarity classes of central separable A-algebras (i.e. Azumaya algebras over A; cf. Azumaya[28]), and many results of the classical theory remain true in this setting. The motivation for this work came from algebraic geometry (cf. Grothendieck[29]), but there are now purely algebraic treatments, such as Knus and Ojanguren[30] and, for a relatively self-contained introduction, Orzech and Small[31].

Skew fields infinite-dimensional over their centre make their first appearance in Hilbert[32], where they illustrate the fact that general ordered skew fields (unlike Archimedean-ordered skew fields) need not be commutative. They have figured in isolated examples (Koethe[33]), as well as in the construction by Ore[34] of the skew field of fractions of a ring with 'Ore condition'. In Cohn[35] an analogue for skew fields of the well-known Higman–Neumann–Neumann construction for groups is obtained, and this is used by Macintyre[36] to construct a skew field with unsolvable word problem, and by Hirschfeld and Wheeler[37], who exhibit second-order models of arithmetic in an existentially closed skew field, and also obtain a non-commutative form of the Nullstellensatz. For a preliminary account of techniques in this area see Cohn[38]; cf. also Cohn[39] for a general criterion for the embeddability of a ring in a skew field, and Cohn[40] for a programme of 'non-commutative algebraic geometry'.

In a slightly different direction the arithmetic in simple algebras leads to the theory of maximal orders. This originated in the study of group rings; if $\mathbf{Q}G$ is the group algebra over $\mathbf{Q}$, the subring $\mathbf{Z}G$ is an order, though not generally maximal. The theory was formulated by Brandt, who showed that the ideals form a groupoid (the Brandt groupoid) under multiplication; cf. Deuring[22] and Jacobson[41]. More recently the theory has been generalised and studied from a homological point of view, in terms of 'hereditary orders', by Auslander and Goldman[42], Brumer[43] and Harada[44]; cf. also Reiner[25].

Group rings have also been studied for infinite groups; here the aim has been to relate properties of the group to those of the ring (Hall[45]). A comprehensive account is in Passman[46], brought up to date in Passman[47], but one of the basic questions remains open: Can the group ring (over $\mathbf{Z}$) of a torsion-free group have zero-divisors? Notable results in this direction were obtained by Lewin and Lewin[48], who prove that the group ring of a torsion-free group with one defining relation is embeddable in a skew field, and Farkas and Snider[49], who prove the same for torsion-free polycyclic groups.

While the Wedderburn theory for algebras was extended at an early stage to Artinian rings (Artin[50]), no such extension to Noetherian rings was to be expected. The correct substitute was found by Goldie[51], who uses the Ore construction, suitably generalised (cf. Asano[52]) to show that a Noetherian (semi-) prime ring has a ring of fractions which is Artinian (semi-) simple. This key result has given rise to a great deal of activity in Noetherian rings; for expositions of various aspects see Lambek[53] and Faith[54]. An interesting study of non-commutative Dedekind rings was made by Robson[55].

The Ore construction and its generalisations was taken up by Gabriel[56] to give a categorical treatment of localisation, leading to torsion theories for Abelian categories and more particularly rings of quotients. A useful survey was given by Stenström[57] (see also Golan[58]), but the theory has probably not yet reached its definitive form. Other important contributions have been made by Lambek[59-61], Goldman[62] and Walker and Walker[63]. For related work in a different direction see Goodearl, Handelman and Lawrence[64].

Rings satisfying polynomial identities, briefly PI-rings, first arose in the study of the foundations of geometry; cf. Dehn[65] and Wagner[66]. Later developments showed a certain parallel to Noetherian rings, with Posner's theorem corresponding to Goldie's theorem. The subject has been developed in a series of papers by Kaplansky, Levitzki, Amitsur, Procesi and others. For a connected account see Procesi[67]. A notable recent result is the construction of central polynomials for matrix rings by Formanek[68] and, independently, Razmyslov[69, 70]. Apart from their intrinsic interest, they have led to simple proofs of Posner's theorem (Rowen[71]) and other results. Azumaya algebras have been characterised by polynomial identities by Artin[72]; his proof has been simplified by Amitsur[73], Procesi[67] and Goldie[74]; see also Cohn[12].

Analysts have studied the ring of linear differential operators and proved a unique factorisation property for it. This went through successive generalisations until Ore[75] treated formal differential equations and skew polynomial rings, by the Euclidean algorithm. Cohn[76] observed that a weaker form of the algorithm holds for a wide class of rings, including free algebras, and is strong enough to allow similar conclusions to be derived. An exposition can be found in Cohn[39], which also includes Bergman's centraliser theorem (Bergman[77]) and Jategaonkar's iterated skew polynomial ring (Jategaonkar[78]).

The structure of Weyl algebras (which occur in quantum mechanics) and enveloping algebras of Lie algebras has given rise to much activity, whose results are collected in Dixmier[79]; see also Gelfand and Kirillov[80]. Recently there has been renewed interest in rings of differential operators through the work of Bernstein[81]; see also Björk[82]. Differential fields have also been studied from the point of view of the functions rather than the operators. This was initiated by Ritt[83]; more recent results are due to Kolchin and others and are summarised in Kolchin[84]. See also Kaplansky[85] and, for the related topic of difference algebra, R.M. Cohn[86]. Closely related to free algebras is the study of free products or 'coproducts' (Cohn[87,88]). Bergman[89] proves some very powerful structure theorems on modules over coproducts of rings, which in effect allow one to construct a ring in which the monoid of finitely generated projectives has a prescribed structure.

From a categorical point of view a K-algebra can be defined as a module A with two operations, multiplication $A \times A \to A$ and unit-element $K \to A$, such that certain diagrams commute. If all arrows are reversed, one obtains a *coalgebra*. A module which has both an algebra and a coalgebra structure that are compatible is called a *Hopf algebra* (since Hopf used this structure on the cohomology ring of a Lie group). A group algebra has a natural Hopf algebra structure, as well as the associative envelope of a Lie algebra, and Hopf algebras occur naturally in many other contexts; for an introduction see Sweedler[90]. Hopf algebras also promise to be of importance in the study of inseparable field extensions; cf. Chase[91], Sweedler[92].

8.3 COMMUTATIVE RINGS

The term 'ring' first occurs in Hilbert's 'Zahlbericht'[93], where it refers to the ring of integers in an algebraic number field. The ideal theory of such rings had been described by Dedekind[94], after whom they have been called. In algebraic geometry a more general kind of ring was needed (Dedekind rings are only 'one-dimensional') and for this purpose the theory of polynomial rings was developed in the early years of this century, but was soon formulated in the more general context of 'Noetherian' rings by Noether[95]. There followed a vigorous development by Noether, Krull, Prüfer, Fitting and others, which is summarised in Krull[96]. Meanwhile the algebraicisation of geometry proceeds apace, and Zariski[97] gives an algebraic treatment of various geometric notions. In particular, the concept of a regular local ring (introduced by Krull[98] under the name p-Reihenring, and developed further by Chevalley[99] and Cohen[100]) is shown to correspond quite generally to a simple point of an algebraic variety. These rings are also characterised homologically and shown to be unique factorisation domains (Serre[101], Auslander and Buchsbaum[102]), and a number of technical improvements, notably the Artin–Rees lemma (Rees[103]), enable the study of completions to be carried further in the general (Noetherian) case.

Many accounts in book form have appeared in the last 15-20 years. Northcott[104] gives a brief but readable introduction; for a very full, indeed leisurely, treatment of the elements (including field theory) going as far as regular local rings, see Zariski and Samuel[105]. Nagata[106] covers rather more (including Henselisation and power series rings, as well as some ingenious counter-examples), but in a more concise form. The lecture notes by Serre[107] give an algebraic treatment of intersection multiplicities. Kaplansky[108] gives a very smooth development of selected parts (no completion) with many substantial exercises. Atiyah and Macdonald[109] is an introductory account with an excellent treatment of dimension theory, and with exercises to stress the geometric analogy. On a larger scale, though still elementary, is Bourbaki[110], which in 7 chapters (over 700 pages and 500 exercises) goes from flat modules and localisation to Krull rings. In 1960 Grothendieck's 'Éléments de géométrie algébrique' began to appear, which has had a very profound influence on the development of commutative algebra. Perhaps the main novelty was that homological methods were used to a greater extent than before and in a geometrical environment this led to progress which, when translated back into algebra, gave new insights into commutative ring theory, but the complexity of the apparatus made this translation a slow process.

8.4 CATEGORIES AND HOMOLOGICAL ALGEBRA

Homological algebra developed from the efforts to give an invariant description of homology groups, in a series of papers by Eilenberg and MacLane, as well as papers by Hurewicz, Hochschild and others. The basic reference is Cartan and Eilenberg[111]; in spite of much subsequent progress this still remains an excellent reference book. A very clear concise introduction can be found in the first chapter of Godement[112]. A larger work stressing the geometrical motivation is MacLane[113], and a more recent treatment emphasising group extensions is Hilton and Stammbach[114].

Category theory, the language underlying homological algebra, at first developed as part of the latter, but is gradually becoming a separate subject, with its own results and problems. One of the basic papers is Grothendieck[115]; here the theory of Abelian categories is developed for use in sheaf theory. A brief readable introduction is Freyd[116]; a somewhat fuller treatment is Mitchell[117]. MacLane[118] is a modern account (though not quite as elementary as the title suggests), Bucur and Deleanu[119] is more oriented towards applications in ring theory, and Herrlich and Strecker[120] is a very broad treatment of fundamentals, with a 50-page bibliography. Some notions that are not really algebraic, such as completion, have been studied in the setting of categories (cf., for example, Lambek[59]) and categories have been used in the study of the foundations of set theory, notably by Lawvere[121].

A basic concept, appearing in many different guises, is that of a *universal construction*. It may be subsumed under the notion of *adjoint functor*; conditions for its existence are described in the adjoint functor theorem (cf. Freyd[116] and MacLane[118]). It gives rise to the notion of *triple* (also called *monad*), which includes not merely universal constructions in algebra, but also other cases such as the Stone–Čech compactification; cf. the symposium edited by Eckmann[122].

Homological algebra has had a profound effect on ring theory. For any ring R, the category $\mathscr{M}_R$ of right R-modules is an Abelian category and categorical notions such as projective or injective module and global dimension have provided basic invariants (Kaplansky[123], Bass[124]). The injective hull is a useful module construction (though in most cases not exactly 'constructive'), which is at the basis of torsion theories for rings, as well as the Matlis–Gabriel decomposition for modules over Noetherian rings (Matlis[125] and Gabriel[56,126]). The dual notion of projective cover need not exist; precise conditions for its existence have been studied by Bass[127], and have led to the notion of perfect and semiperfect ring; cf. Björk[128]. Another interesting study of the lack of symmetry of $\mathscr{M}_R$ is by Chase[129]. The homological dimension of rings and modules is examined in a long series of papers in the 1950s in the *Nagoya* and *Pacific Journals of Mathematics*. In particular, *hereditary* rings (i.e. of global dimension at most 1) have been studied from many points of view; cf. for example, Small[130], Jategaonkar[78] and Bergman[131]. For a way in which set-theoretic questions enter into the discussion, see Osofsky[132]. The global dimension of rings of linear differential operators, and more generally of Weyl algebras, has been examined by Rinehart[133], Roos[134] and Björk[135]. The notion of dimension has been applied to Abelian categories themselves, taking $\mathscr{M}_R$ to be 0-dimensional and making a 'modular resolution' by categories of the form $\mathscr{M}_R$; cf. Roos[134].

Morita[136] made a detailed study of duality between module categories. This led Chase, Schanuel, Bass and others to examine the case of equivalence between module categories, and it gave rise to the notion of *Morita equivalence* of rings, which has become a most useful classification principle; cf. Bass[137] and Cohn[138] and, for a detailed account, Bass[139,140].

It has only lately been realised that linear algebra can in principle be done for modules over any arbitrary ring, once the correct obstruction groups have been recognised. This has led to the development of *K*-theory (for which the impetus came from algebraic topology and geometry), and among the first successes were the stability theorems proved by Bass[141] and Bass, Heller and Swan[142], using the model from algebraic geometry provided by Serre[143]. A more detailed treatment, making contact with many parts of algebra and number theory, and giving an informative review of categorical algebra for the ring theorist, is Bass[140]. An earlier set of notes, Bass[139], deals in more detail with the case of quadratic forms. All these references concentrate on the K_0 and K_1 groups; it was only after prolonged experimentation that a satisfactory form of K_2 and higher K-groups was found. This was mainly the work of Quillen, building on the results by Bass, Swan, Gersten, Villamayor and others, and is described in the Proceedings of the Battelle Institute Conference (Bass[144]). The K_2-group has also been studied for global fields by Bass and Tate; see Tate[145] and, for a connected account, Milnor[146].

8.5 LIE ALGEBRAS AND OTHER NON-ASSOCIATIVE ALGEBRAS

Lie algebras first arose as the rings of linear differential operators in a Lie group (i.e. continuous group). The best-known example is the space of 3-dimensional vectors with the operation of vector multiplication. This is just the Lie algebra of the 3-dimensional rotation group. The classification of semisimple Lie algebras was begun by Killing and completed by E. Cartan[147] in his thesis. The series of papers by Weyl[144] on representations of semisimple Lie groups is most readable and still of interest. The modern standard account of Lie algebra theory is Jacobson[149]; the actual classification is in terms of Coxeter–Dynkin diagrams, which also occur in many other contexts (cf. Coxeter[150] for a classification of finite rotation groups). Much work has been done on representations of Lie algebras, partly with a view to physical applications. The connection of Lie algebras with finite groups is studied by Zassenhaus[151] and is carried further by Lazard[152]. For an account of the analogy with infinite group theory see Amayo and Stewart[153].

In the study of formal Lie groups the Lie algebra is less prominent, since it is not enough to determine the Lie algebra in characteristic $p \neq 0$, the case of chief interest. Its place is taken by a certain Hopf algebra (also called bigebra); cf. Dieudonné[154], also Manin[155].

From a formal point of view Jordan algebras are parallel to Lie algebras; in fact they are not as closely linked to associative algebras as are Lie algebras, but they seem to have just as rich a theory. A thorough account is Jacobson[156]; another point of view which relates them more closely to algebraic groups (and uses the classification theory of the latter) is adopted by Springer[157]. Vinberg[158] uses Jordan algebras in the study of convex domains, and (apart from the original applications, to quantum mechanics) Jordan algebras and Jordan triple systems also enter into the study of differential geometry; cf. Loos[159]. For an account of the exceptional Jordan algebra and the connection with Lie algebras, see Jacobson[160].

The best-known alternative algebra is the Cayley–Dickson algebra (octonions), formed from the quaternions in the same way as the latter were formed from the complex numbers. Their special role is highlighted by the Bruck–Kleinfeld theorem (Bruck and Kleinfeld[161] and Skomyakov[162]). Dorofeev[163] has obtained results which show that free alternative rings behave very differently from either free associative or free Lie rings.

Besides these algebras there are many other types of non-associative algebras, usually defined by identities, which have been studied at various times; cf. Albert[164] and Schafer[165]; for a recent survey see McCrimmon[166].

A different generalisation of a ring is obtained by giving up one distributive law and the commutative law of addition. The resulting structure is a *near-ring*; it arises naturally as the set of all mappings of a non-Abelian group into itself. In the last two decades there has been a flourishing theory of near-rings, which in many ways parallels the theory of rings, but the most important case is that of *near-fields*, whose use in the study of projective planes goes back to Dickson. The classification of finite near-fields (Zassenhaus[167]) has been used by Amitsur[168] to determine all finite subgroups of skew fields. For a recent survey of near-fields, see Wähling[169].

8.6 MONOIDS AND MORE GENERAL SYSTEMS

To study multiplication in its purest form one naturally turns to *semigroups* (systems with an associative multiplication) or *monoids* (semigroups with 1). Monoids are not nearly of the same importance as groups, but with the greater penetration of the whole field they have come to be studied quite intensively in recent years and there are several accounts of the subject, e.g. Lyapin[170] and, for a survey of the whole field, Clifford and Preston[171]. The main structure theory is due to Suschkewitsch, Rees, Green and Schützenberger. A readable introduction (also taking recent work into account) can be found in Howie[172].

Monoids play a role in coding theory (a comma-free code is essentially a free set in a free monoid), and much work has been done by Schützenberger and his school. Algebraic language theory and automata theory also have close links with monoids; cf. Gross and

Lentin[173]. A comprehensive account which generalises and unifies much of what is known is Eilenberg[174]; for a brief survey of language theory see Cohn[175]. Many of the papers in the journal *Information and Control* deal with these topics. The automata aspects are stressed in Arbib[176]. A study of monoids from this point of view (including the notion of 'complexity' of a monoid) has been made by Krohn and Rhodes[177].

Recently biologists have studied models of cell division (Lindenmayer systems) which closely resemble context-free languages, and this development has raised interesting questions about endomorphisms in free monoids; cf. Herman and Rozenberg[178] and Rozenberg and Salomaa[179].

Of the many other systems with a binary operation that have been studied (and are surveyed in Bruck[180]), *loops* and *quasi-groups* are the most important after monoids. They arise in geometry (Blaschke and Bol[181]) and in connection with alternative fields, but the main application is to cubics and cubic surfaces. The addition on a non-singular cubic curve is usually described in terms of an Abelian group, but this group is really derived from a quasi-group which arises naturally. In the same way a cubic hypersurface gives rise to a quasi-group which is associated with a commutative Moufang loop. A very lucid account of the connection can be found in Manin[182].

8.7   PERIODICALS

The spate of activity in algebra in recent years is reflected by the fact that there are now several periodicals devoted exclusively to algebraic topics:

Algebra i Logika (Akademiya Nauk SSSR. Sibirskoe Otdelenie, Novosibirsk, 1962-   ); vol. 7 onwards translated as *Algebra and Logic* (Consultants Bureau, New York, 1968-   )

Communications in Algebra (Dekker, New York, 1974-   )

Information and Control (Academic Press, New York, 1958-   )

* *Journal of Algebra* (Academic Press, New York, 1964-   )

* *Journal of Pure and Applied Algebra* (North-Holland, Amsterdam, 1972-   )

Semigroup Forum (Springer, Berlin, 1972-   )

Many recent conference and seminar reports have appeared in the Springer Lecture Note series (cf. Section 2.9). In particular, there have been many numbers devoted to category theory.

---

* Fortunately these two do not (yet) live entirely up to their name: they only appear monthly and quarterly, respectively.

The *Proceedings* of the International Congress of Mathematicians are a good source of survey articles. The Congress takes place every four years, e.g. Vancouver, 1974, and the proceedings appear a year or two later. At a somewhat lower level, but often very informative, are the survey articles in the *American Mathematical Monthly* and in *Elemente der Mathematik*. The surveys in the *Bulletin* of the London Mathematical Society and the selected addresses in the *Bulletin* of the American Mathematical Society are usually very readable, although sometimes more technical. The *Russian Mathematical Surveys* are also useful, although here there is a preponderance of analysis.

REFERENCES

The list of works referred to in the text is followed by a list, in alphabetical order of author, of other works closely related to it.

1. Jacobson, N., *Lectures in abstract algebra, Vol. II*, Van Nostrand, Princeton (1953)
2. Herstein, I.N., *Topics in algebra*, Blaisdell, Boston, Mass. (1964)
3. Lang, S., *Algebra*, Addison-Wesley, Reading, Mass. (1965)
4. Cohn, P.M., *Algebra, Vol. I*, Wiley, London (1974)
5. Bernstein, I.N., Gelfand, I.M. and Ponomarev, V.A., 'Coxeter functors and Gabriel's theorem', *Uspekhi Matematicheskikh Nauk*, **28**, 19-33 (1973); translated in *Russian Mathematical Surveys*, **28**, 17-32 (1973)
6. Gelfand, I.M. and Ponomarev, V.A., 'Problems of linear algebra and classification of quadruples of subspaces in a finite-dimensional vector space', *Colloquia Mathematica Societatis János Bolyai. Proceedings*, **5**, 163-237 (1970)
7. Gabriel, P., 'Unzerlegbare Darstellungen I', *Manuscripta Mathematica*, **6**, 71-103 (1972)
8. Gabriel, P., 'Indecomposable representations II', *Symposia Mathematica*, **11**, 81-104 (1973)
9. Dlab, V. and Ringel, C.M., *Representations of graphs and algebras*, Carleton Lecture Notes, Carleton University, Ottawa (1974)
10. Ringel, C.M., 'Representations of *K*-species and bimodules', *Journal of Algebra*, [to appear]
11. Curtis, C.R. and Reiner, I., *Representations theory of finite groups and associative algebras*, Wiley, New York (1962)
12. Cohn, P.M., *Algebra, Vol. II*, Wiley, London (1977)
13. Roiter, A.V., 'Unboundedness of the dimensions of indecomposable representations of algebras having infinitely many indecomposable representations', *Akademiya Nauk SSSR. Izvestiya–Seriya Matematicheskaya*, **32**, 1275-1282 (1968)
14. Steinitz, E., 'Algebraische Theorie der Körper', *Journal für die Reine und Angewandte Mathematik*, **137**, 167-308 (1910); reprinted Teubner, Leipzig (1930)
15. Jacobson, N., *Structure of rings*, Colloquium Publications 37, American Mathematical Society, Providence, R.I. (rev. edn., 1956)

16. Chevalley, C., *Introduction to the theory of algebraic functions of one variable,* Mathematical Surveys 6, American Mathematical Society, New York (1951)
17. Šafarevič, I.R., *Basic algebraic geometry,* Springer, Berlin (1974)
18. Lang, S., *Diophantine geometry,* Wiley, New York (1962)
19. Weber, H., *Lehrbuch der Algebra,* Teubner, Leipzig (1908); reprinted Chelsea, New York (1965)
20. Lenstra, H.W., 'Rational functions invariant under a finite abelian group', *Inventiones Mathematicae,* **25**, 299-325 (1974)
21. Chase, S.U., Harrison, D.K. and Rosenberg, A., 'Galois theory and Galois cohomology of commutative rings', *American Mathematical Society. Memoirs,* **52**, 1-79 (1965)
22. Deuring, M., *Algebren,* Springer, Berlin (1934); reprinted Chelsea, New York (1948)
23. Serre, J.-P., *Corps locaux,* Hermann, Paris (1962)
24. Weil, A., *Basic number theory,* Springer, Berlin (1967)
25. Reiner, I., *Maximal orders,* London Mathematical Society Monographs 5, Academic Press, London (1975)
26. Cassels, J.W.S. and Fröhlich, A., *Algebraic number theory,* Academic Press, London (1967)
27. Auslander, M. and Goldman, O., 'The Brauer group of a commutative ring', *American Mathematical Society. Transactions,* **97**, 367-409 (1960)
28. Azumaya, G., 'On maximally central algebras', *Nagoya Mathematical Journal,* **2**, 119-150 (1951)
29. Grothendieck, A., 'Le groupe de Brauer, I-III, Séminaire Bourbaki 1965/66' in *Dix exposés sur la cohomologie des schémas* (ed. J. Giraud et al. North-Holland, Amsterdam (1968)
30. Knus, M.A. and Ojanguren, M., *Théorie de la descente et algèbres d'Azumaya,* Lecture Notes in Mathematics 389, Springer, Berlin (1973)
31. Orzech, M. and Small, C., *The Brauer group of commutative rings,* Dekker, New York (1975)
32. Hilbert, D., *Über die Grundlagen der Geometrie,* Teubner, Leipzig (1898)
33. Koethe, G., 'Schiefkörper unendlichen Ranges über dem Zentrum', *Mathematische Annalen,* **105**, 15-39 (1931)
34. Ore, O., 'Linear equations in non-commutative fields', *Annals of Mathematics,* **32**, 463-477 (1931)
35. Cohn, P.M., 'On the embedding of firs in skew fields', *London Mathematical Society. Proceedings,* **23**, 193-213 (1971)
36. Macintyre, A., 'Combinatorial problems for skew fields' [to appear]
37. Hirschfeld, J. and Wheeler, W.H., *Forcing, arithmetic, division rings,* Lecture Notes in Mathematics 454, Springer, Berlin (1975)
38. Cohn, P.M., *Skew field constructions,* London Mathematical Society Lecture Note Series 27, Cambridge University Press, Cambridge (1977)
39. Cohn, P.M., *Free rings and their relations,* London Mathematical Society Monographs 2, Academic Press, London (1971)
40. Cohn, P.M., 'Progress in free associative algebras', *Israel Journal of Mathematics,* **19**, 109-151 (1974)
41. Jacobson, N., *Theory of rings,* Mathematical Surveys 2, American Mathematical Society, New York (1943)
42. Auslander, M. and Goldman, O., 'Maximal orders', *American Mathematical Society. Transactions,* **97**, 1-24 (1960)
43. Brumer, A., 'Pseudocompact algebras, profinite groups and classformations', *Journal of Algebra,* **4**, 442-470 (1966)

44. Harada, M., 'Structure of hereditary orders', *Osaka City University. Journal of Mathematics*, 14, 1-22 (1963)
45. Hall, P., 'On the finiteness of certain soluble groups', *London Mathematical Society. Proceedings*, 9, 595-622 (1959)
46. Passman, D.S., *Infinite group rings*, Dekker, New York (1971)
47. Passman, D.S., 'Advances in group rings', *Israel Journal of Mathematics*, 19, 67-107 (1974)
48. Lewin, J. and Lewin, T., 'An embedding of the group algebra of a torsionfree 1-relator group in a field', *Journal of Algebra* [to appear]
49. Farkas, D.R. and Snider, R.L., 'K_0 and Noetherian group rings', *Journal of Algebra*, 42, 192-198 (1976)
50. Artin, E., 'Zur Theorie der hyperkomplexen Zahlen', *Hamburg, Universität: Mathematische Seminar. Abhandlungen*, 5, 251-260 (1928); reprinted in *Collected papers*, Addison-Wesley, Reading, Mass. (1965)
51. Goldie, A.W., 'The structure of rings under ascending chain conditions', *London Mathematical Society. Proceedings*, 8, 589-608 (1958)
52. Asano, K., 'Über die Quotientenbildung von Schiefringen', *Mathematical Society of Japan. Journal*, 1, 73-78 (1949)
53. Lambek, J., *Rings and modules*, Blaisdell, Boston, Mass. (1966)
54. Faith, C., *Algebra: rings, modules and categories I*, Springer, Berlin (1973)
55. Robson, J.C., 'Non-commutative Dedekind rings', *Journal of Algebra*, 9, 249-265 (1968)
56. Gabriel, P., 'Des catégories abéliennes', *Société Mathématique de France. Bulletin*, 90, 323-448 (1962)
57. Stenström, B., *Rings and modules of quotients*, Lecture Notes in Mathematics 237, Springer, Berlin (1971)
58. Golan, J.S., *Localization of noncommutative rings*, Dekker, New York (1975)
59. Lambek, J., *Completion of categories*, Lecture Notes in Mathematics 24, Springer, Berlin (1966)
60. Lambek, J., *Torsion theories, additive semantics and rings of quotients*, Lecture Notes in Mathematics 177, Springer, Berlin (1971)
61. Lambek, J., 'Noncommutative localization', *American Mathematical Society. Bulletin*, 79, 857-872 (1973)
62. Goldman, O., 'Rings and modules of quotients', *Journal of Algebra*, 13, 10-47 (1969)
63. Walker, E.A. and Walker, C.L., 'Quotient categories and rings of quotients', *Rocky Mountain Journal of Mathematics*, 2, 513-555 (1972)
64. Goodearl, K.R., Handelman, D. and Lawrence, J., *Strongly prime and completely torsionfree rings*, Carleton Mathematical Series 109, Carleton University, Ottawa (1974)
65. Dehn, M., 'Über die Grundlagen der projektiven Geometrie und allgemeine Zahlensysteme', *Mathematische Annalen*, 85, 184-193 (1922)
66. Wagner, W., 'Über die Grundlagen der projektiven Geometrie und allgemeine Zahlensysteme', *Mathematische Annalen*, 113, 528-567 (1936-1937)
67. Procesi, C., *Rings with polynomial identities*, Dekker, New York (1973)
68. Formanek, E., 'Central polynomials for matrix rings', *Journal of Algebra*, 23, 129-133 (1972)
69. Razmyslov, Yu. P., 'On a problem of Kaplansky', *Akademiya Nauk SSSR. Izvestiya–Seriya Matematicheskaya*, 37, 483-501 (1973)
70. Razmyslov, Yu. P., 'Trace identities on full matrix rings over fields of characteristic 0', *Akademiya Nauk SSSR. Izvestiya–Seriya Matematicheskaya*, 38, 723-756 (1974)

71. Rowen, L.H., 'On rings with a central polynomial', *Journal of Algebra*, 31, 393-426 (1974)
72. Artin, M., 'On Azumaya algebras and finite-dimensional representations', *Journal of Algebra*, 11, 532-563 (1969)
73. Amitsur, S.A., 'Polynomial identities and Azumaya algebras', *Journal of Algebra*, 27, 117-125 (1973)
74. Goldie, A.W., 'Azumaya algebras and rings with polynomial identity', *Cambridge Philosophical Society. Proceedings*, 79, 393-399 (1976)
75. Ore, O., 'Theory of non-commutative polynomials', *Annals of Mathematics*, 34, 480-508 (1933)
76. Cohn, P.M., 'On a generalization of the Euclidean algorithm', *Cambridge Philosophical Society. Proceedings*, 57, 18-30 (1961)
77. Bergman, G.M., 'Centralizers in free associative algebras', *American Mathematical Society. Transactions*, 137, 327-344 (1969)
78. Jategaonkar, A.V., 'A counter-example in homological algebra and ring theory', *Journal of Algebra*, 12, 418-440 (1969)
79. Dixmier, J., *Algèbres enveloppantes*, Gauthier-Villars, Paris (1974)
80. Gelfand, I.M. and Kirillov, A.A., 'Sur les corps liés aux algèbres enveloppantes des algèbres de Lie', *Institut des Hautes Études Scientifiques. Publications mathématiques*, 31, 5-19 (1966)
81. Bernstein, I.N., 'The analytic continuation of generalized functions with respect to a parameter', *Funktsional'nyĭ Analiz I Ego Prilozheniya*, 6, 26-40 (1972); translated in *Functional Analysis and its Applications*, 6, 273-285 (1973)
82. Björk, J.-E., 'I.N. Bernstein's functional equation' [to appear]
83. Ritt, J.F., *Differential algebra*, Colloquium Publications 33, American Mathematical Society, New York (1950)
84. Kolchin, E.R., *Differential algebra and algebraic groups*, Academic Press, New York (1973)
85. Kaplansky, I., *An introduction to differential algebra*, Hermann, Paris (1955)
86. Cohn, R.M., *Difference algebra*, Wiley, New York (1965)
87. Cohn, P.M., 'On the free product of associative rings, I', *Mathematische Zeitschrift*, 71, 380-398 (1959)
88. Cohn, P.M., 'On the free product of associative rings, II', *Mathematische Zeitschrift*, 73, 433-456 (1960)
89. Bergman, G.M., 'Modules over coproducts of rings', *American Mathematical Society. Transactions*, 200, 1-32 (1974)
90. Sweedler, M.E., *Hopf algebras*, Benjamin, New York (1969)
91. Chase, S.U., 'On inseparable Galois theory', *American Mathematical Society. Bulletin*, 77, 413-417 (1971)
92. Sweedler, M.E., 'The predual theorem to the Jacobson Bourbaki correspondence', *American Mathematical Society. Transactions*, 213, 391-406 (1975)
93. Hilbert, D., 'Die Theorie der algebraischen Zahlkörper', *Deutsche Mathematiker-Vereinigung. Jahresbericht*, 4, 175-546 (1897)
94. Dedekind, R., 'Über die Theorie der ganzen algebraischen Zahlen', 11th supplement to *Vorlesungen über Zahlentheorie* (P.G. Lejeune Dirichlet), reprinted Vieweg, Braunschweig (1964)
95. Noether, E., 'Idealtheorie in Ringbereichen', *Mathematische Annalen*, 83, 24-66 (1921)
96. Krull, W., *Idealtheorie*, Springer, Berlin (1935); reprinted Chelsea, New York (1948)

97. Zariski, O., 'The concept of a simple point of an abstract algebraic variety', *American Mathematical Society. Transactions*, **62**, 1-52 (1947); reprinted in *Collected papers*, Vol. 1, M.I.T. Press, Cambridge, Mass. (1972)

98. Krull, W., 'Dimensionstheorie in Stellenringen', *Journal für die Reine und Angewandte Mathematik*, **179**, 204-226 (1938)

99. Chevalley, C., 'On the theory of local rings', *Annals of Mathematics*, **44**, 690-708 (1943)

100. Cohen, I.S., 'On the structure and ideal theory of complete local rings', *American Mathematical Society. Transactions*, **59**, 252-261 (1946)

101. Serre, J.-P., 'Sur la dimension homologique den anneaux et des modules Noethériens', in *Symposium on algebraic number theory*, Tokyo-Nikko (1955)

102. Auslander, M. and Buchsbaum, D., 'Unique factorization in regular local rings', *National Academy of Sciences, United States. Proceedings*, **45**, 733-734 (1959)

103. Rees, D., 'Two classical theorems of ideal theory', *Cambridge Philosophical Society. Proceedings*, **52**, 155-157 (1956)

104. Northcott, D.G., *Ideal theory*, Cambridge University Press, Cambridge (1953)

105. Zariski, O. and Samuel, P., *Commutative algebra*, 2 vols, Van Nostrand, Princeton (1958, 1960)

106. Nagata, M., *Local rings*, Wiley, New York (1962)

107. Serre, J.-P., *Algèbre locale, multiplicités*, Lecture Notes in Mathematics 11, Springer, Berlin (1965)

108. Kaplansky, I., *Commutative rings*, Allyn and Bacon, Boston, Mass. (1970)

109. Atiyah, M.F. and Macdonald, I.G., *Introduction to commutative algebra*, Addison-Wesley, Reading, Mass. (1969)

110. Bourbaki, N., *Algèbre commutative*, 4 vols, Hermann, Paris (1961-1965)

111. Cartan, H. and Eilenberg, S., *Homological algebra*, Princeton University Press, Princeton (1956)

112. Godement, R., *Théorie des faisceaux*, Hermann, Paris (1958)

113. MacLane, S., *Homology*, Springer, Berlin (1963)

114. Hilton, P.J. and Stammbach, U., *A course in homological algebra*, Springer, Berlin (1971)

115. Grothendieck, A., 'Sur quelques points d'algèbre homologique', *Tôhoku Mathematical Journal*, **9**, 119-221 (1957)

116. Freyd, P., *Abelian categories, an introduction to the theory of functors*, Harper and Row, New York (1964)

117. Mitchell, B., *Theory of categories*, Academic Press, New York (1965)

118. MacLane, S., *Categories for the working mathematician*, Springer, Berlin (1971)

119. Bucur, I. and Deleanu, A., *Categories and functors*, Wiley, New York (1968)

120. Herrlich, H. and Strecker, G.E., *Category theory*, Allyn and Bacon, Boston, Mass. (1973)

121. Lawvere, F.W., 'The category of categories as a foundation for mathematics' in *Proceedings of the Conference on categorical algebra, La Jolla, California, 1965* (ed. by S. Eilenberg *et al.*). Springer, Berlin (1966)

122. Eckmann, B., *Seminar on triples and categorical homology theory*, Lecture Notes in Mathematics 80, Springer, Berlin (1969)

123. Kaplansky, I., 'Projective modules', *Annals of Mathematics*, **68**, 372-377 (1958)

124. Bass, H., 'Big projective modules are free', *Illinois Journal of Mathematics*, **7**, 24-31 (1963)

125. Matlis, E., 'Injective modules over Noetherian rings', *Pacific Journal of Mathematics*, **8**, 511-528 (1958)

126. Gabriel, P., 'Objets injectifs dans les catégories abéliennes', *Séminaire Dubreil-Pisot 1958/59,* Institut Henri Poincaré, Paris (1959)

127. Bass, H., 'Finitistic dimension and a homological generalization of semi-primary rings', *American Mathematical Society. Transactions,* **95,** 466-488 (1960)

128. Björk, J.-E., 'Rings satisfying a minimum condition on principal ideals', *Journal für die Reine und Angewandte Mathematik,* **236,** 112-119 (1969)

129. Chase, S.U., 'Direct products of modules', *American Mathematical Society. Transactions,* **97,** 457-473 (1960)

130. Small, L.W., 'Orders in semihereditary rings', *American Mathematical Society. Bulletin,* **73,** 656-658 (1967)

131. Bergman, G.M., 'Hereditary commutative rings and the centers of hereditary rings', *London Mathematical Society. Proceedings,* **23,** 214-236 (1971)

132. Osofsky, B.L., 'Homological dimension and the continuum hypothesis', *American Mathematical Society. Transactions,* **132,** 217-230 (1968)

133. Rinehart, G.S., 'Note on the global dimension of a certain ring', *American Mathematical Society. Proceedings,* **13,** 341-346 (1962)

134. Roos, J.-E., 'On the structure of abelian categories with generators and exact direct limits' [to appear]

135. Björk, J.-E., 'Global dimension of algebras of differential operators', *Inventiones Mathematicae,* **17,** 69-78 (1972)

136. Morita, K., 'Duality for modules and its application to the theory of rings with minimum condition', *Tokyo Kyoiku Daigaku. Science Reports. A,* **6,** 83-142 (1958)

137. Bass, H., *The Morita theorems,* Lecture notes, University of Oregon, Eugene (1962)

138. Cohn, P.M., *Morita equivalence and duality,* Lecture notes, Queen Mary College, University of London (1966; reprinted 1974)

139. Bass, H., *Topics in algebraic K-theory,* Tata Institute of Fundamental Research, Bombay (1967)

140. Bass, H., *Algebraic K-theory,* Benjamin, New York (1968)

141. Bass, H., 'K-theory and stable algebra', *Institut des Hautes Études Scientifiques. Publications Mathématiques,* **22,** 5-60 (1964)

142. Bass, H., Heller, A. and Swan, R.G., 'The Whitehead group of a polynomial extension', *Institut des Hautes Études Scientifiques. Publications Mathématiques,* **22,** 61-79 (1964)

143. Serre, J.-P., 'Modules projectifs et espaces fibrés à fibre vectorielle', *Séminaire Dubreil 23, 1957/58,* Institut Henri Poincaré, Paris (1958)

144. Bass, H. (ed.), *Algebraic K-theory,* 3 vols, Lecture Notes in Mathematics 341, 342, 343, Springer, Berlin (1973)

145. Tate, J.T., 'Symbols in arithmetic', *International Congress of Mathematicians. Proceedings,* **1,** 201-211 (1971)

146. Milnor, J.W., *Introduction to algebraic K-theory,* Princeton University Press, Princeton, N.J. (1971)

147. Cartan, E., 'Sur la structure des groupes de transformations finis et continus', in *Oeuvres complètes,* 6 vols, Gauthier-Villars, Paris (1952-1955)

148. Weyl, H., 'Theorie der Darstellung kontinuierlicher halbeinfacher Gruppen durch lineare Transformationen', *Mathematische Zeitschrift,* **23,** 271-309 (1925); **24,** 328-376, 377-395 (1926)

149. Jacobson, N., *Lie algebras,* Wiley, New York (1962)

150. Coxeter, H.S.M., *Regular polytopes,* Macmillan, London (1948)

151. Zassenhaus, H., 'Ein Verfahren, jeder endlichen p-Gruppe einen Lie Ring mit der Charakteristik p zuzuordnen', *Hamburg, Universität: Mathematisches Seminar. Abhandlungen,* **13,** 200-207 (1939)

152. Lazard, M., 'Sur des groupes nilpotents et les anneaux de Lie', *École Normale Supérieure. Annales Scientifiques,* 3me serie, 71, 101-190 (1954)
153. Amayo, R. and Stewart, I.N., *Infinite-dimensional Lie algebras,* North-Holland, Amsterdam (1974)
154. Dieudonné, J., *Introduction to the theory of formal groups,* Dekker, New York (1973)
155. Manin, Yu. I., 'The theory of formal commutative groups over fields of finite characteristic', *Uspekhi Matematicheskikh Nauk,* 18(6), 3-90 (1963); translated in *Russian Mathematical Surveys,* 18, 1-83 (1963)
156. Jacobson, N., *Structure and representation of Jordan algebras,* Colloquium Publications 39, American Mathematical Society, Providence, R.I. (1968)
157. Springer, T.A., *Jordan algebras and algebraic groups,* Springer, Berlin (1973)
158. Vinberg, E.B., 'The theory of homogeneous convex cones', *Moskovskoe Matematicheskoe Obshchestvo. Trudy* 12, 303-358 (1963); translated in *Moscow Mathematical Society. Transactions,* 12, 340-403 (1963)
159. Loos, O., *Lectures on Jordan triples,* University of British Columbia, Vancouver, B.C. (1974)
160. Jacobson, N., *Exceptional Lie algebras,* Dekker, New York (1967)
161. Bruck, R.H. and Kleinfeld, E., 'The structure of alternative division rings', *American Mathematical Society. Proceedings,* 2, 878-890 (1951)
162. Skornyakov, L.A., 'Alternative fields', *Ukrainskiĭ Matematicheskiĭ Zhurnal,* 2, 70-85 (1950) [in Russian]
163. Dorofeev, G.V., 'Alternative rings on three generators', *Sibirskiĭ Matematicheskiĭ Zhurnal,* 4, 1029-1048 (1963) [in Russian]
164. Albert, A.A., 'Power-associative rings', *American Mathematical Society. Transactions,* 64, 552-593 (1948)
165. Schafer, R.D., *An introduction to non-associative algebras,* Academic Press, New York (1966)
166. McCrimmon, K., 'Quadratic methods in non-associative algebras', *International Congress of Mathematicians. Proceedings,* 1, 325-330 (1975)
167. Zassenhaus, H., Über endliche Fastkörper', *Hamburg, Universität. Mathematisches Seminar. Abhandlungen,* 11, 187-220 (1936)
168. Amitsur, S.A., 'Finite subgroups of division rings', *American Mathematical Society. Transactions,* 80, 361-386 (1955)
169. Wähling, H., 'Bericht über Fastkörper', *Deutsche Mathematiker-Vereinigung. Jahresbericht,* 76, 41-103 (1974)
170. Lyapin, E.S., *Semigroups,* 2nd edn, Translations of Mathematical Monographs 3, American Mathematical Society, Providence, R.I. (1968)
171. Clifford, A.H. and Preston, G.B., *The algebraic theory of semigroups,* 2 vols, Mathematical Surveys 7, American Mathematical Society, Providence, R.I. (1961, 1967)
172. Howie, J.M., *Introduction to semigroup theory,* London Mathematical Society Monographs 7, Academic Press, London (1976)
173. Gross, M. and Lentin, A., *Introduction to formal grammars,* Springer, Berlin (1970)
174. Eilenberg, S., *Automata, languages and machines,* vol. A, Academic Press, New York (1974)
175. Cohn, P.M., 'Algebra and language theory', *London Mathematical Society. Bulletin,* 7, 1-29 (1975)
176. Arbib, M.A. (ed.), *Algebraic theory of machines, languages and semigroups,* Academic Press, New York (1968)
177. Krohn, K. and Rhodes, J., 'Complexity of finite semigroups', *Annals of Mathematics,* 88, 128-160 (1968)

178. Herman, G.T. and Rozenberg, G., *Developmental systems and languages,* North-Holland, Amsterdam (1974)
179. Rozenberg, G. and Salomaa, A., *L-systems,* Lecture Notes in Computer Science 15, Springer, Berlin (1974)
180. Bruck, R.H., *A survey of binary systems,* Springer, Berlin (1958)
181. Blaschke, W. and Bol, G., *Geometrie der Gewebe,* Springer, Berlin (1938)
182. Manin, Yu. I., *Kubicheskie formy: algebra, geometriya, arifmetika,* Nauka, Moscow (1972); translated as *Cubic forms: algebra, geometry, arithmetic,* North-Holland, Amsterdam (1974)

FURTHER READING

Albert, A.A., *Structure of algebras,* Colloquium Publications 24, American Mathematical Society, New York (1939)

Amitsur, S.A., 'Rational identities and applications to algebra and geometry', *Journal of Algebra,* 3, 304-359 (1966)

Amitsur, S.A., 'On central division algebras', *Israel Journal of Mathematics,* 12, 408-420 (1972)

Amitsur, S.A., 'Polynomial identities', *Israel Journal of Mathematics,* 19, 183-199 (1974)

Bergman, G.M., 'Coproducts and some universal ring constructions', *American Mathematical Society. Transactions,* 200, 33-88 (1974)

Bernstein, I.N., 'Modules over a ring of differential operators', *Funktsional'nyǐ Analiz i ego Prilozheniya,* 5, 1-16 (1971); translated in *Functional Analysis and its Applications,* 5, 89-101 (1971)

Bokut, L.A., 'Factorization theorems for certain classes of rings without zero-divisors', *Algebra i Logika,* 4(4), 25-52; 4(5), 17-46 (1965) [in Russian]

Bokut, L.A., 'On Mal'cev's problem', *Sibirskiǐ Matematicheskiǐ Zhurnal,* 10, 965-1005 (1969) [in Russian]

Buchsbaum, D. and Eisenbud, D., 'What makes a complex exact?', *Journal of Algebra,* 25, 259-268 (1973)

Buchsbaum, D. and Eisenbud, D., 'Some structure theorems for finite-free resolutions', *Advances in Mathematics,* 12, 84-139 (1974)

Dlab, V. and Ringel, C.M., 'On algebras of finite representations type', *Journal of Algebra,* 33, 306-394 (1975)

Dorofeev, G.V., 'An example in the theory of alternative rings', *Sibirskiǐ Matematicheskiǐ Zhurnal,* 4, 1049-1052 (1963)

Eisenbud, D. and Evans, E.G., 'Generating modules efficiently: theorems from algebraic *K*-theory', *Journal of Algebra,* 27, 278-305 (1973)

Goodearl, K.R., *Nonsingular rings and modules,* Dekker, New York (1976)

Gordon, R. and Robson, J.C., 'Krull dimension', *American Mathematical Society. Memoirs,* 13, (1973)

Herstein, I.N., *Non-commutative rings,* Wiley, New York (1968)

Hochster, M., 'The equicharacteristic case of some homological conjectures in local rings' [to appear]

Jacobson, N., *Basic algebra, Vol I,* Freeman, San Francisco (1974)

Kaplansky, I., 'Problems in the theory of rings revisited', *American Mathematical Monthly,* 77, 445-454 (1970)

Kleinfeld, E., 'A characterization of the Cayley numbers' in *Studies in modern algebra,* Vol. 2 (ed. A.A. Albert), Prentice-Hall, Englewood Cliffs, N.J. (1963)

Krohn, K., Langer, R. and Rhodes, J., 'Algebraic principles for the analysis of a biochemical system', *Journal of Computer and Systems Sciences,* **1,** 119-136 (1967)

Kurke, H., Pfister, G. and Roczen, M., *Henselsche Ringe und Algebraische Geometrie,* Deutscher Verlag der Wissenschaften, Berlin (1975)

Peskine, C. and Szpiro, L., 'Dimension projective finie et cohomologie locale', *Institut des Hautes Études Scientifiques. Publications Mathématiques,* **42,** (1973)

Pilz, G., *Near-rings,* Mathematical Studies 23, North-Holland, Amsterdam (1977)

Procesi, C., 'The invariant theory of $n \otimes n$ matrices' [to appear]

Quillen, D., 'Projective modules over polynomial rings', *Inventiones Mathematicae,* **36,** 167-171 (1976)

Regev, A., 'Existence of polynomial identities in A B, *Israel Journal of Mathematics,* **11,** 131-151 (1972)

Stenström, B., *Rings of quotients,* Springer, Berlin (1975)

Swan, R.G., 'Vector bundles and projective modules', *American Mathematical Society. Transactions,* **105,** 264-277 (1962)

Swan, R.G., *Algebraic K-theory,* Lecture Notes in Mathematics 76, Springer, Berlin (1968)

9

Group Theory

Peter M. Neumann

The theory of groups and its close relatives include such a wide range of mathematics that we find it necessary to make some subdivision in order to be able to describe its present state. Accordingly our account is organised under the following six headings: introductory works; finite groups; representation theory; infinite groups; algebraic and Lie groups; and related structures. As in any such subject, the various parts merge into one another in a continuous pattern, and any partitioning is necessarily subjective and unimportant. Our choice of division does not indicate an equality of weight or importance measured by volume of publication or any other such parameter. It is based on a subjective feeling that each of these areas has its own distinctive style.

9.1 INTRODUCTORY WORKS

In the bibliography that ends this chapter we have described as 'text' some books appropriate as introduction to the subject, which do not concentrate on one particular aspect of it. They all treat the axiomatics of group theory, subgroups, cosets and the so-called Lagrange's Theorem, normal subgroups and quotient groups, and homomorphisms. Nearly all treat direct products, extensions, the structure theorem for finite Abelian groups, Sylow's Theorems for finite groups, and the presentation of groups by generators and relations. As much as this can also be found in very many of the textbooks on abstract algebra that are written for the use of undergraduates or beginning postgraduate students.

The books by Ledermann[1], Macdonald[2], Rotman[3] and Schmidt[4] are suitably elementary for undergraduates. The remaining textbooks are appropriate for advanced undergraduates or for postgraduate students, and each has its distinguishing points. The beautifully written work of Kurosh[5,6] is devoted mainly to infinite groups; Marshall Hall[7] includes material on permutation groups, on the Burnside Problem, and on applications to the study of projective planes, which has been unusually influential; the book by Hall[7], Schenkman[8], Scott[9] and Speiser[10] contain chapters providing an introduction to the representation theory of finite groups; Hall[7] and Scott[9] also offer a description of the so-called transfer homomorphism, which is not readily accessible elsewhere at this level and is a useful tool in finite group theory. Speiser's book[10] contains a beautiful treatment of the symmetry of ornaments.

9.2 FINITE GROUPS

There has been spectacular progress made in these last two decades in the study of finite groups. Much of this has been concerned with or inspired by the search for finite simple groups, as comparison of the bibliography by Davis[11] with the collection of reviews edited by Gorenstein[12] shows. Following Galois (1832), a group is said to be simple if, apart from itself and the trivial group, it has no normal subgroups. Since a finite group G having a non trivial normal subgroup H may be thought of as being built from the smaller groups G/H and H, each of which may be further analysed in this way, every finite group may ultimately be considered as being constructed from simple groups much as every integer can be obtained as a product of prime numbers. (The analogy goes a little further in that the Jordan–Hölder Theorem may be considered to be a sort of unique factorisation theorem. It guarantees that the simple components of a group are independent of the manner in which it is analysed.)

Apart from cyclic groups of prime order, the best-known simple groups are probably the alternating groups A_n (for $n \geqslant 5$) and the projective special linear groups $PSL(n,q)$ that are defined as the factor groups $SL(n,q)/Z$, where $SL(n,q)$ denotes the group of n by n matrices of determinant 1 with entries from the Galois field of q elements, and Z denotes the group of those scalar matrices which it contains. Similar constructions with orthogonal matrices, symplectic matrices and unitary matrices over suitable fields give further families of simple groups that were known already last century and are known as the 'classical' groups (see Jordan[13], Dickson[14] and Dieudonné[15]). In an enormously influential article published in 1955 Chevalley[16] showed how most of these and similar groups could be obtained in a uniform way as automorphism groups of analogues of the classical simple Lie algebras defined over finite fields. His construction was supplemented by a process of 'twisting' introduced and exploited by Steinberg, Ree and Tits, and by 1962 all the known infinite families of simple groups, including the Suzuki groups that had just been discovered as the solution of a problem concerning permutation groups, were accounted for. These Chevalley groups and twisted versions, generally known as groups of 'Lie type' are described in a monograph by Carter[17].

Apart from these infinite families only finitely many simple groups, the so-called 'sporadic' groups are known. Five of these, the Mathieu groups M_{11}, M_{12}, M_{22}, M_{23}, M_{24}, have been known for over 100 years. The others have been discovered at irregular intervals since 1965. The last chapter of Carter[17] is devoted to sporadic groups; more exhaustive information is to be found in the survey article by Feit[18]. To bring his list up to date (1976) one should add

R of order	$2^{14}.3^3.5^3.7.13.29$	('Rudvalis')
ONS	$2^9.3^4.5.7^3.11.19.31$	('O'Nan–Sims')
F $(=F_5)$	$2^{14}.3^6.5^6.7.11.19$	('Harada–Norton')
E $(=F_3)$	$2^{15}.3^{10}.5^3.7^2.13.19.31$	('Thompson–Smith')
B? $(=F_2)$	$2^{41}.3^{13}.5^6.7^2.11.13.17.19.23.31.47$	('Baby Monster')
M? $(=F_1)$	$2^{46}.3^{20}.5^9.7^6.11^2.13^3.17.19.23.29.31.41.47.59.71$	
		('Fischer's Monster')
Ja?	$2^{21}.3^3.5.7.11^3.23.29.31.37.43$	('New Janko')

Of these the first four have been proved to exist: the Rudvalis group by Conway and Wales[19], the O'Nan–Sims group by Sims in 1973 (see O'Nan[20]), the Thompson–Smith group by P.E. Smith[21] and the Harada–Norton group by S. Norton in his PhD thesis (Cambridge, 1974; cf. Harada[22]). A construction of the Baby Monster using machine computation was announced by J.S. Leon and C.C. Sims in November, 1976 (unpublished). See also Fischer[22a]. The remaining two groups have been described by Griess[23] and Fischer, and Janko[23a], respectively, but their existence is not yet (1976) established. It is typical of this area that one may analyse the structure of such a possibility down to the smallest detail and still not know of its existence as a certainty, only as a likelihood that grows stronger as more detail is tabulated without any inconsistency appearing.

These new sporadic simple groups have been found by many methods. Some have arisen from the development of the theory, some by chance as automorphism groups of interesting geometric or combinatorial structures, some are the results of experimental searches, and some have been proved to exist by machine computation. In view of the large variety of apparently accidental circumstances that allow the existence of these new groups, it seems entirely possible that a substantial number of them is still be discovered.

While the search for new examples is continuing, the general analysis and classification of simple groups is being prosecuted with even more vigour. Indeed, the discovery of some of the new groups was hardly more than an exciting by-product of these theoretical investigations. The most famous breakthrough was the theorem of Feit and Thompson[24], published in 1963, that the only simple groups of odd order are the cyclic groups of prime order. Since then the classification of simple groups in terms of the structure of their Sylow 2-subgroups, or of the centralisers of elements of order 2, or of other internal properties, has had great success, not only in the results obtained, but also in the development of new techniques. Thus, for example, all simple groups whose Sylow 2-subgroups are Abelian or of nilpotency class two are known (Walter[25], Bender[26], Gilman and Gorenstein[27]); so are simple groups in which the Sylow 2-subgroup contains a non-trivial strongly-closed Abelian subgroup (Goldschmidt[28]); and recent work of Aschbacher, Goldschmidt, Thompson and Foote has both developed 'signaliser functor theorems' that appear to be widely applicable and also promised the classification of all groups containing an element of order 2 whose centraliser is not 2-constrained.

The book by Gorenstein[29] explains most of the general background information about finite groups and most of the basic techniques, except for signaliser functors, that are used in these classification theorems. The articles by Glauberman, Gorenstein and Dade in the *Proceedings*[30] of a conference in Oxford, 1969, and by Gorenstein and Walter[31], are expositions going rather deeper. Surveys have been written by Feit[18] and Gorenstein[32]. The former has a good bibliography, although because of the rapid development of the subject parts of it are already a little out of date; the latter has no bibliography and is less comprehensive, but provides a delightfully readable account of the aims, the tactics and the strategy of current research.

One of the areas of finite group theory that is in some ways quite close to the search for simple groups is the study of permutation groups. This lay dormant for many years, but has recently sprung to life again, revitalised partly by the writings of Wielandt[33, 34] partly by the discovery of some of the new finite simple groups as interesting permutation groups. Several of these were in fact discovered as groups of rank 3: the 'rank' of a group G of permutations of the set Ω is defined to be the number of orbits of G acting on the set Ω^2 of all ordered pairs from Ω. For groups of rank 3, of which there are many interesting examples among the naturally occurring geometrically defined groups, there is a rich theory that has been developed principally by D.G. Higman. If the order of the rank 3 group G is even (which in practice is always the case), then G has exactly two orbits on the set $\Omega^{\{2\}}$ of unordered pairs from Ω. Taking one of these orbits as the edge set one obtains a graph with vertex set Ω admitting G as a group of automorphisms. This graph has many virtues. It is, for example, 'strongly regular', and the eigenspaces of its adjacency matrix correspond to the reduction of the permutation character of G. Generally, it is the representation of the group by automorphisms of the graph and of closely related combinatorial structures that provides the facility for understanding the group. Although the theory is at its best for groups of rank 3, it works well also for groups of higher rank. One finds that each orbit of G on Ω^2 gives rise to a graph. It is the structure of each graph, the relationship between the different graphs, and the structure of the algebra spanned by the adjacency matrices of the graphs (known as the 'centraliser algebra' of G, or, sometimes, as the 'Hecke algebra' because of the similarity to the Hecke ring of an arithmetic subgroup of a Lie group), that provide the means for studying G. The theory of rank 3 groups is surveyed by Higman[35]; an elementary exposition of the graph theory and centraliser ring theory in general is given by Neumann[36]; a deeper survey is given by Cameron[37]; a very general study of the combinatorial theory involved has been made by Higman[38, 39].

In the preceding paragraph our permutation groups were of rank 3 or more. A group is of rank 2 precisely when it is multiply transitive. Just as simply transitive groups may be studied as automorphism groups of graphs, so doubly transitive groups frequently act as automorphism groups of geometric or combinatorial structures like block designs or Steiner systems. A survey of this theory, with an excellent bibliography, has been given by Kantor[40]. The techniques and the successes of recent work on simple groups have been applied also to the study of doubly transitive groups by Aschbacher, Bender, Holt, O'Nan and others. As a result for example, the normal structure of the point-stabiliser of a doubly transitive group is essentially known (O'Nan[41]), and all doubly transitive groups in which the point stabilisers are soluble are known (O'Nan[42] and Holt[43]). One might hope that, since high transitivity is so very much stronger than double transitivity, the new theorems should yield the long-sought classification of fourfold or fivefold transitive groups, or at least an (affirmative) answer to the 100-year-old question of Jordan, whether the alternating and symmetric groups are the only sixfold transitive groups. However, the new techniques are at their best when a one-point stabiliser has a non-trivial Abelian normal subgroup, and this almost never happens in a highly transitive group. Quite probably the classical questions of permutation group theory, of which Jordan's is the most famous, will have to wait for their solution until the classification of finite simple groups is completed, when answers will be obtainable by simply checking through lists.

A third part of permutation group theory that has made significant progress recently concerns the old search for transitive groups of prime degree p, and newer questions of a similar kind about groups of degree $2p$, $3p$, ..., p^2, p^3, ... Here the main tool is representation theory. Both ordinary characters and modular representations have been used with success. The survey and the lecture notes by Neumann[44, 45] concentrate mainly on groups of prime degree and have extensive bibliographies; Wielandt[34] introduces important and interesting new techniques of wide applicability in this area.

As mentioned at the beginning of this section, one way to analyse a finite group is to split it into smaller ones, a normal subgroup and its factor group, and thus ultimately to arrive at simple constituents, known as its composition factors. For this idea to be useful it is of course necessary that one should understand the reverse process to that of forming the factor group by a normal subgroup. What is required is machinery for describing all groups G having a given group A as normal subgroup with a specified group B as factor group G/A. By using techniques based on the Sylow Theorems and the so-called Frattini Argument, one may usually reduce to the case where the normal subgroup A is nilpotent, and then indeed an Abelian p-group for some prime number p. At this point the methods of homological algebra (see p. 145) are available to give a survey of the possible extensions. Elementary expositions of extension theory are to be found in, for example, Kurosh[5] and Rotman[3]; more extensive accounts in Hall[7], Huppert[46], Zassenhaus[47] and MacLane[48].

The theory of finite soluble groups is a good illustration of the philosophy indicated in the preceding paragraph. A group is said to be soluble precisely when its composition factors are cyclic groups of prime orders. In a famous paper in 1928 Hall[49] used induction on the group order and information about extensions of groups to prove an important generalisation of the Sylow Theorems for soluble groups. By similar methods, in a similarly celebrated paper in 1961, Carter[50] showed the existence of a unique conjugacy class of self-normalising nilpotent subgroups and related them to the chief factors in a finite soluble group. An elegant theory has followed, developed by P. Hall, Carter, Gaschütz, Hawkes and others. It is explained in Chapter VI of Huppert[46], and is the subject of a promised monograph by Doerk and Hawkes.

Since most of the theory of finite groups depends heavily on the Sylow Theorems, which guarantee the existence of subgroups of prime-power order in any group, the study of these, the so-called *p*-groups, is indispensable for an understanding of the general theory. However, because questions arise from all parts of finite group theory, problems concerning *p*-groups are too varied to be surveyed here. The study of *p*-groups goes back to Sylow and Burnside, but the modern theory is based on another famous paper of Hall[51], where, *inter alia*, the commutator calculus was introduced. There is an exposition, for example, in Hall[7] and an extended account in Huppert[46] (Chapter III). The associated Lie ring, which in effect linearises commutator calculations, is described in Gorenstein[29], in Lazard[52] and in Magnus, Karrass and Solitar[53]. A catalogue of the 2-groups of orders up to 64, with an interesting account of the appropriate theory, is given by Hall and Senior[54].

9.3 REPRESENTATION THEORY

Representation theory is usually taken to be the art and science and technique of portraying groups as groups of matrices, or, equivalently, as groups of linear transformations of finite-dimensional vector spaces over a field F. It applies to finite groups on the one hand; Lie groups and algebraic groups on the other. Although these two parts of mathematics have similar foundations, their flavours and their further ramifications are quite different: the representation theory of finite groups is strongly arithmetical and ring-theoretical; the representation theory of Lie groups and algebraic groups becomes closely connected with Fourier analysis and other parts of functional analysis. We will limit ourselves to finite groups here, and to the algebraic side of the representation theory of Lie groups briefly in Section 9.5.

Let G be a finite group. A representation of G of degree n over the field F is a homomorphism $\rho:G \to GL(n,F)$, where $GL(n,F)$ is the group of invertible (i.e. non-singular) n by n matrices over F. The representations ρ_1, ρ_2 are said to be equivalent, and are treated as indistinguishable, if there is an invertible n by n matrix T such that $T^{-1} g\rho_1 T = g\rho_2$ for all g in G. Associated with a representation ρ is a module for the group algebra FG, and, conversely, a (unital) module for the ring FG, finite-dimensional over F, gives rise in a natural way to a representation of G; equivalence of representations is the same thing then as isomorphism of modules; in this way the techniques of ring theory may be brought to bear. If the field F is of characteristic 0, typically the complex number field or a sufficiently large algebraic number field, then one speaks of 'ordinary' representations. When the characteristic of F is a prime number p not dividing the order $|G|$, the theory turns out to be almost exactly the same as the ordinary representation theory for *G*. If the characteristic of *F* is a prime number *p* that divides $|G|$, then one speaks of 'modular' or '*p*-modular' representations.

The ordinary representations of *G* are determined by their 'characters', the character associated with ρ being that function χ on *G* to *F* whose value at *g* is the trace of the matrix $g\rho$. Consequently, ordinary representation theory is usually treated as character theory. Every character can be expressed as a sum of the so-called 'irreducible' characters, of which *G* has only finitely many, and so the ordinary representations of *G* are known when one has its character table. This is a convenient display of the values of the (absolutely) irreducible characters of *G*. Since the values of the characters are always algebraic integers (lying in a suitable cyclotomic number field), and since they must satisfy many conditions (such as Schur's orthogonality relations, or that the restriction to a subgroup is a character), it is often possible to deduce the character table, or a significant fragment of it, from a very small amount of information about the internal structure of *G*. Then, in turn, knowledge of the characters of *G* yields further information about its structure. Used in this way, character theory is an important tool in the search for simple groups and in permutation group theory.

The earliest textbook exposition is that in Burnside[55], which is still of interest (in spite of the out-of-date language used), not only for historical reasons, but also because of the applications that Burnside gives. Further introductory expositions are to be found in Hall[7], Schenkman[8], Speiser[10], Gorenstein[29] and Serre[56], among others. More extensive treatments are given by Huppert[46], Curtis and Reiner[57], Dornhoff[58], Feit[59] and Isaacs[60].

Modular representation theory is significantly more complicated than the ordinary theory. We owe it mainly to the work of Richard Brauer, who developed it as a tool for finding and classifying the ordinary characters of *G* and for proving deep arithmetical properties of them. His bridge between ordinary and modular representations is, in effect, the so-called 'integral' representations: if *R* is a sub-ring of the field *F* of characteristic zero, such that *F* is the field of fractions of *R* and the prime *p* is not invertible in *R*, then a representation of *G* by matrices over *R* on the one hand can be viewed as a representation ρ of *G* over *F*, on the other produces a representation over a field *K* of characteristic *p* when the entries of the matrices are all reduced modulo a maximal ideal *M* of *R* that contains *p*. For elementary accounts of this process see Curtis and Reiner[57] (Chapter XII), Serre[56] or Dornhoff[58] (Part B). The article by Dade in the *Proceedings*[30] exhibits beautifully how the theory may be directly applied to the search for simple groups. For deeper results it is still best to consult the many original articles of Brauer, and of others such as Feit, Dade and

J.A. Green. In particular, Green's philosophy seems to be a little different from that of Brauer. He studies the modular representations of G, without reference to characters, using mildly homological methods. This lends itself well to ring-theoretic exposition such as Green[61] and Hamernik[62] give. An exhaustive exposition of all aspects of the theory is given in lecture notes by Feit[63].

The p-modular representation theory of a group whose Sylow p-subgroups are cyclic is particularly well worked out and has found many applications. It was originally produced by Brauer and has since been simplified by Thompson[64] and extended by Dade[65] and many others. Accounts are to be found in Dornhoff[58], Green[61] and Feit[63]. This theory is applied almost every time that the character table of a large simple group is computed (see, for example, O'Nan[20] and Hall[66]), and it has also proved useful in the study of certain permutation groups (see Neumann[45] for exposition and references).

One of the aims of representation theory is the explicit calculation of the character tables of all finite groups. This was begun by Frobenius for the symmetric and alternating groups, and there are many expositions of his results now available in works by Littlewood[67], Robinson[68], Coleman[69], Kerber[70] and others. The calculation of the characters of general linear groups over finite fields was accomplished in 1955 by Green[71], and recent work of Carter and Lusztig[72], Deligne and Lusztig[73] and others (see articles in Borel *et al.*[74], and the survey by Springer[75]) represents good progress towards the production of both the ordinary and the modular characters of the remaining finite groups of Lie type. Character tables of the sporadic simple groups are known, and indeed have frequently played a large part in their discovery, but have not yet been collected in one published work. The bibliography of Feit[18] gives sources for many of these groups; tables for the smaller simple groups were collected by McKay and Lambert[76]; a complete collection is being compiled by J.H. Conway and R.T. Curtis in Cambridge, but is not yet generally available.

9.4 INFINITE GROUPS

As in the collection of reviews edited by Baumslag[77], the theory of infinite groups is to mean the general study of groups, in which finiteness or topologies play no special part. The work by Kurosh[5] still remains the best introduction to this area. A supplement added to the third Russian edition[6] published in 1967 contains a survey and a rich bibliography covering the years 1952–1965. There is some hope that this will be translated and published as a third volume of the American edition[5]. Other textbooks that contain introductions to parts of the subject, usually free groups and Abelian groups, are those by Ledermann[1], Macdonald[2], Rotman[3], Hall[7], Schenkman[8] and Scott[9].

One of the earliest and strongest reasons for studying infinite groups arose from the roles they play in analysis and topology, in the theory of automorphic functions, as monodromy groups of linear differential equations, and as fundamental groups of manifolds and other topological spaces. In such contexts a group will usually be first described by a presentation in terms of generators and relations, and a large part of the theory of infinite groups is concerned with these presentations. The monographs by Coxeter and Moser[78], by Magnus, Karrass and Solitar[53] and by Lyndon and Schupp[79] offer detailed exposition and good bibliographies. The last of these contains an account of group presentations in which relatively small amounts of cancellation are possible between different relators, treated by the geometrical and topological methods introduced by Lyndon which have all but superseded the older 'sieve' method of Tartakovskii. It also contains an account of 'bipolar structures' in free products. These were first introduced by Stallings in 1968 in his proof that groups of cohomological dimension one are free, and an account in the context of topological motivations and applications is to be found in his monograph[80]. An elegant technique of a similar type, that of groups acting on trees, due to Serre and Bass[81], is available in lecture notes which may appear in the Springer Lecture Notes in Mathematics series.

Most of the groups that arise in practice in analysis or topology are finitely presented — that is, presented in terms of finitely many generators and finitely many relations. For finite presentations some of the most important questions are algorithmic: the so-called 'word problem' (whether there is an algorithm to decide of any given product of the generators and their inverses whether or not it is the unit element of the group), the 'conjugacy problem', the 'isomorphism problem' and others. Examples of finitely presented groups with insoluble word problem were given already over 20 years ago by Novikov, Britton and Boone. On the other hand algorithms for solving the word problem are known for a variety of important classes of finitely presented groups such as one-relator groups, suitably small cancellation groups, residually finite groups and simple groups, and efforts to widen the scope of results of this kind are continuing. Not nearly so much is known yet in regard to the conjugacy problem and the isomorphism problem. Even for one-relator presentations, for which the solution of the word problem was achieved by Magnus in 1932, it is still not known whether these other problems are soluble or not. This general area has a strongly logical flavour as one uses recursion theory to prove the non-existence of algorithms and the undecidability of certain questions. The monograph by Miller[82] besides containing new results, contains a useful survey of these logical problems. Baumslag has written a more general survey[83], containing a good bibliography, of finitely presented groups, and a specialist survey[84] of problems about one-relator groups. An old and influential question, known as the Burnside Problem, may be viewed as a question about generators and relations. In its original version it was essentially the question whether a finitely generated group, all of whose elements are of finite order, must itself be finite. A counter-example was first produced by Golod[85] in 1964; an interesting and entertainingly different approach was exhibited by Alešin[86] in 1972. However, in the course of time, the Burnside problem for exponent n has come to mean the question whether a finitely generated group, all of whose elements are of finite order dividing n, is necessarily finite. This is known to be true if n is 1, 2, 3, 4 or 6 (cf. Hall[7], Chapter 18) and is now known to be false if n is a multiple of an odd number not less than 665. A solution was published by Novikov and Adjan in 1968, and improved in Adjan[87]; another solution, announced in 1964, was given by Britton[88].

Most of the interesting variants of the Burnside problem, and many questions that are related to it, are of a varietal nature. A variety of groups is a class of groups that can be axiomatised by adding further identical relations to the axioms describing groups in general. Thus, for example, the class of all Abelian (commutative) groups is a variety and may be defined by the identical relation $xy = yx$, as also is the class $\mathscr{B}_n$, the Burnside variety consisting of all groups of exponent dividing n, which may be defined by the relation $x^n = 1$. The study of varieties in general properly belongs to model theory and universal algebra (see, for example, Cohn[89]), but it is in group theory that varieties have arisen most naturally, and it is varieties of groups that have been most intensively studied. The monograph by Neumann[90] gives an excellent survey of this subject. It is already somewhat out of date because of the progress made on the problems formulated in it; an article by Kovács and Newman[91] gives relatively recent information on those problems and an extension of its bibliography.

Some parts of infinite group theory have arisen by analogy or generalisation from finite-dimensional Lie groups or from finite groups. Thus, an infinite group is said to be nilpotent if it has a central series of finite length, or to be soluble if it has a normal series of finite length with Abelian factors. The commutator calculus originated by P. Hall[51], and its reflection in properties of an associated Lie ring, is as valuable for the study of nilpotent groups in general as for finite p-groups (see p. 140). Much is known in particular about finitely generated nilpotent groups. The notes by Hall[92] and Baumslag[93] are excellent sources. Methods for treating infinite soluble groups were described by Hall in a trilogy[94-96] which is still the best introduction to the subject. There is a vast collection of other properties that have been suggested and studied as generalisations of nilpotence or solubility, generalisations in the sense that one of these is what they reduce to in finite groups. Kurosh[5] (Chapters XIII, XIV, XV) provides a good introduction to this area; the monograph by Robinson[97], supplemented by his article[98], gives an exhaustive recent survey with a huge bibliography. In another direction, a generalisation of finiteness that has received considerable attention, particularly with reference to generalised Sylow theorems and to the study of simple groups, is that of local finiteness. A group is said to be locally finite if every finite set of elements is contained in a finite subgroup. The monograph by Kegel and Wehrfritz[99] offers a good introduction to the subject, a good survey and an equally good bibliography.

Many of the infinite groups that arise in analysis and geometry are linear — that is to say, groups of invertible n by n matrices over a field. Although some results go back to Burnside and Schur in the first decade of this century, the first systematic work on linear groups was done by Mal'cev in 1940. An exposition of the theory in a very wide context is given by Plotkin[100], a less ambitious exposition by Dixon[101], and a deeper survey with a good bibliography by Wehrfritz[102].

Group-rings are also receiving considerable attention at present from specialists both in group theory and in ring theory. The study of infinite soluble groups leads directly into questions about group-rings, as Hall[94-96] showed, and the subject has also received considerable impetus from efforts to prove that there are no zero-divisors in the group-ring of a torsion-free group and that the Jacobson radical of a group algebra over a field of characteristic zero is trivial. The monograph by Passman[103] is already a little out of date, probably because of the influence of the list of research problems in its final chapter, but he has published a more recent introductory article[104] which is a useful supplement.

In its 30 years' history the cohomology theory of groups has developed to a sizeable area of algebra, with applications not only in the algebraic topology and extension theory (see p. 139) from which it arose, but also in number theory, in geometry and in representation theory. The first book on the subject, an enormously influential one, was that of Cartan and Eilenberg[105]; but any good textbook on homological algebra in general, such as those by MacLane[48] or Hilton and Stammbach[106], will include an easier introduction to the homology and cohomology theory of groups. The books by Kurosh[5], Rotman[3], Hall[7] and Huppert[46] all contain a certain amount of the theory developed as a tool for use in treating the extension problem. The cohomology theory of finite groups, partly because finiteness has useful consequences making for a richer theory, partly because of the influence of number-theoretic applications and requirements upon it, has a special flavour. This is apparent in the book by Babakhanian[107], and in the expository article by Atiyah and Wall[108]. The notes by Gruenberg[109] and Stammbach[110], on the other hand, give considerable information about cohomology and homology, particularly for infinite groups. There is a particularly useful chapter (the eighth) in the former on finite cohomological dimension, a topic which is the subject of work in progress at the moment by Bieri, K.F. Brown, Chiswell, Eckmann, Serre and others, who are producing a good theory of Euler characteristic for a wide class of infinite groups. Cohomology groups can be defined in very great generality for so-called 'triplable' categories – in particular, for varieties in the sense of universal algebra. An interesting study of this triple cohomology in varieties of groups has been produced by Leedham-Green[111] and his collaborators.

Perhaps the largest autonomous branch of infinite group theory is the study of Abelian groups. The books by Kurosh[5], Rotman[3], Hall[7] and Schenkman[8] give brief introductions going a little beyond the standard structure theorem for finitely generated Abelian groups. For further exposition, and for some indication of how the subject generalises to modules over principal ideal rings there is the little red book of Kaplansky[112], which, in its second edition, contains useful notes on the literature. The standard reference in this area is the two-volume monograph by Fuchs[113], which supersedes his earlier book (Pergamon Press, Oxford, 1960). However, quite recently part of the subject has been somewhat changed by an invasion of infinite combinatorics and model theory introduced by Shelagh[114], who has shown that the solution of the so-called Whitehead Problem depends upon one's ambient set theory, and who has solved some of the other problems in Fuchs' monograph.

9.5   ALGEBRAIC AND LIE GROUPS

The theories of algebraic groups and of Lie groups are considerably more geometric and topological than algebraic. Nevertheless, there are points of contact and similarity with some parts of the theory of discrete groups, and it is these points which we wish to sketch briefly here.

An algebraic group is a group structure in the category of algebraic varieties – that is, it is a group whose underlying set is an algebraic variety, and in which the multiplication and inverse are given by polynomial functions in terms of suitable local coordinates. Similarly, a Lie group is a group structure in the category of smooth manifolds (or manifolds with real-analytic structure). It turns out that there is a strong resemblance between the theories of algebraic groups and of Lie groups. The main difference is that the former can be defined and studied over any field, even over fields of finite characteristic, and consequently there are correspondingly greater technical difficulties in the theorems and proofs. There are books by Borel[115] and Humphries[116] which offer good introductions to algebraic groups. The former gives few references to other works, but has good bibliographical notes appended to many of its sections. The book by Humphries has an extensive bibliography. There are rather more introductory works on Lie groups available, with points of view ranging widely from differential topology to physics. Two of the best-known are Bourbaki's[117], which is an exhaustive treatment, but whose bibliography is not large, and Chevalley's[118], which is a standard reference, but which has no bibliography.

A finite-dimensional connected algebraic or Lie group has a unique maximal soluble normal subgroup and the quotient group which is 'semisimple' is essentially a direct product of simple groups. Thus, by comparison with finite groups, whose construction from their composition factors may be very complicated indeed, the construction of connected algebraic or Lie groups from 'simple' ones is relatively well understood. Furthermore, the simple ones are explicitly known. For Lie groups this is a classical result of Killing and Cartan; its analogue for algebraic groups was proved by Chevalley, and later by Borel and Tits. The book by Carter[17] explains how they are related to the finite simple groups. In a remarkable set of notes, Tits[119] gives an analysis of the geometry behind the simple algebraic groups sufficient for their classification and for use in constructing the finite groups of Lie type also.

For applications in physics and chemistry and in various branches of pure mathematics it is the representation theory of Lie groups which is important. One of the earliest and most influential works on this topic was that of Weyl[120]. He expounds the relationship (due essentially to Schur) between the representation theory of the symmetric group and the decomposition of the tensor representations of the general linear group (or of the orthogonal group), the relationship between the representations of a connected semisimple group and those of its maximal compact subgroup (Weyl's 'unitary trick'), and the use of integration over a compact group to replace the averaging process that is used when dealing with the characters of a finite group. A more recent book on this subject with a good bibliography is that of Boerner[121]. There is a good introduction to the representation theory of compact groups by Adams[122]. This is rather different from the older treatments. It does not seek to calculate the characters explicitly; on the other hand, its abstract approach and elegant techniques make it an excellent survey of the theory.

9.6 RELATED STRUCTURES

The subjects which come in *Mathematical Reviews* at the end of the section described as 'Group theory and generalisations' are nowadays as easily considered to be universal algebra as variants of group theory. Therefore our mention of them here will be very brief. The monograph by Bruck[123] is still the best source of information on some of these structures, particularly loops. However, the theory of semigroups has now developed a large literature of its own; and literature concerning loops is often to be found classified as combinatorial theory of latin squares.

In the theory of semigroups the two-volume work by Clifford and Preston[124] has been the standard source for several years. The first volume is a textbook account of the structure and representation theory of semigroups stemming from J.A. Green's theory of ideals and the equivalence relations that they give rise to and from a theorem of Rees on a form of matrix representation of semigroups. The second volume contains miscellaneous information about presentations, embeddings, congruences, representations and other topics. The structure of finitely generated commutative semigroups is treated exhaustively by Rédei[125], and a recent monograph by Howie[126] contains *inter alia* a good introduction and a long chapter on inverse semigroups (i.e. semigroups in which every element a has an 'inverse' x such that $xax = x$ and $axa = a$). In the last few years a new and promising line has developed, motivated by applications in the theory of finite state machines. This can be found in context in, for example, the books by Arbib[127] and Eilenberg[128]. The appropriate decomposition theory of semigroups, first used in this way by Krohn and Rhodes, involves the so-called wreath product, which arose first in several parts of group theory. An elementary survey of wreath products, with good bibliographical information, is given by Wells[129], and up-to-date articles on the decomposition theory have been written by Rhodes[130] and his collaborators.

REFERENCES

1. Ledermann, W., *Introduction to group theory,* Oliver and Boyd, Edinburgh (1973) [text]
2. Macdonald, Ian D., *The theory of groups,* Clarendon Press, Oxford (1968) [text]
3. Rotman, Joseph J., *The theory of groups: an introduction,* 2nd edn, Allyn and Bacon, Boston (1973) [text]
4. Schmidt, O.U., *Abstract theory of groups,* Freeman, San Francisco (1966) [text]

5. Kurosh, A.G., *The theory of groups,* 2 vols, Chelsea, New York (1956, 1960); translated from the second Russian edition [text]
6. Kurosh, A.G., *Teoriya grupp,* 3rd edn, Nauka, Moscow (1967)
7. Hall, Marshall, Jr., *The theory of groups,* Macmillan, New York (1959) [text]
8. Schenkman, Eugene, *Group theory,* Van Nostrand, Princeton (1965) [text]
9. Scott, W.R., *Group theory,* Prentice-Hall, Englewood Cliffs, N.J. (1964) [text]
10. Speiser, Andreas, *Die Theorie der Gruppen von endlicher Ordnung,* 4th edn, Birkhäuser, Basel (1956) [text]
11. Davis, Constance, *A Bibliographical survey of simple groups of finite order, 1900-1965,* Courant Institute of Mathematical Sciences, New York University, New York (1969)
12. Gorenstein, Daniel (ed.), *Reviews on finite groups, as printed in Mathematical Reviews, 1940-1970, Volumes 1-40,* American Mathematical Society, Providence, R.I. (1974)
13. Jordan, Camille, *Traité des substitutions et des équations algébriques,* Gauthier-Villars, Paris (1870)
14. Dickson, Leonard Eugene, *Linear groups with an exposition of the Galois field theory,* 2nd edn, Dover, New York (1958)
15. Dieudonné, Jean, *La géométrie des groupes classiques,* 2nd edn, Springer, Berlin (1963)
16. Chevalley, C., 'Sur certain groupes simples', *Tôhoku Mathematical Journal,* 7, 14-66 (1955)
17. Carter, Roger W., *Simple groups of Lie type,* Wiley, London (1972)
18. Feit, Walter, 'The current situation in the theory of finite simple groups', *International Congress of Mathematicians. Proceedings,* 1, 55-93 (1970)
19. Conway, J.H. and Wales, D.B., 'Construction of the Rudvalis group of order 145,926,144,000', *Journal of Algebra,* 27, 538-548 (1973)
20. O'Nan, Michael E., 'Some evidence for the existence of a new simple group', *London Mathematical Society. Proceedings,* 32, 421-479 (1976)
21. Smith, P.E., 'A simple subgroup of M? and $E_8(3)$', *London Mathematical Society. Bulletin,* 8, 161-165 (1976)
22. Harada, Koichiro, 'On the simple group F of order $2^{14}.3^6.5^6.7.11.19$' in *Proceedings of the Conference on finite groups, Utah 1975* (ed. William R. Scott and Fletcher Gross), Academic Press, New York (1976)
22a. Fischer, Bernd, 'Finite groups generated by 3-transpositions', *Inventiones Mathematicae,* 13, 232-246 (1971)
23. Griess, Robert L., Jr., 'The structure of the "Monster" simple group' in *Proceedings of the Conference on finite groups, Utah 1975* (ed. William R. Scott and Fletcher Gross), Academic Press, New York (1976)
23a. Janko, Zvonimir, 'A new finite simple group of order 86.775. 571.046.077.562.880 which possesses M_{24} and the full covering group of M_{22} as subgroups', *Journal of Algebra,* 42, 564-596 (1976)
24. Feit, Walter and Thompson, John G., 'Solvability of groups of odd order', *Pacific Journal of Mathematics,* 13, 775-1029 (1963)
25. Walter, John H., 'The characterization of finite groups with Abelian Sylow 2-subgroups', *Annals of Mathematics,* 89, 405-514 (1969)
26. Bender, Helmut, 'On groups with Abelian Sylow 2-subgroups', *Mathematische Zeitschrift,* 117, 164-176 (1970)
27. Gilman, Robert and Gorenstein, Daniel, 'Finite groups with Sylow 2-subgroups of class 2', *American Mathematical Society. Transactions,* 207, 1-126 (1975)

28. Goldschmidt, David M., '2-fusion in finite groups', *Annals of Mathematics*, 99, 70-117 (1974)
29. Gorenstein, Daniel, *Finite groups*, Harper and Row, New York (1968)
30. Powell, M.G. and Higman, G. (eds), *Finite simple groups. Proceedings of a NATO–LMS instructional conference, Oxford 1969*, Academic Press, London (1971)
31. Gorenstein, Daniel and Walter, John H., 'Balance and generation in finite groups', *Journal of Algebra*, 33, 224-287 (1975)
32. Gorenstein, Daniel, 'Finite simple groups and their classification', *Israel Journal of Mathematics*, 19, 5-66 (1974)
33. Wielandt, Helmut, *Finite permutation groups*, Academic Press, New York (1964)
34. Wielandt, Helmut, *Permutation groups through invariant relations and invariant functions*, Lecture Notes, Ohio State University, Columbus, Ohio (1969)
35. Higman, D.G., 'A survey of some questions and results about rank 3 permutation groups', *International Congress of Mathematicians. Proceedings*, 1, 361-365 (1970)
36. Neumann, Peter M., 'Finite permutation groups, edge-coloured graphs and matrices', in *Proceedings of the conference on group theory and computation, Galway 1973*, Academic Press, London (1977)
37. Cameron, P.J., 'Suborbits in transitive permutation groups' in *Combinatorics* (ed. M. Hall, Jr. and J.H. van Lint)., Nato Advanced Study Institute (Series C) 16, Reidel, Dordrecht (1974)
38. Higman, D.G., 'Coherent configurations, I', *Padova, Universita degli Studi di: Seminario Matematico. Rendiconti*, 44, 1-25 (1970)
39. Higman, D.G., *Combinatorial considerations about permutation groups*, Lecture Notes, Mathematical Institute, Oxford (1972)
40. Kantor, W.M., '2-transitive designs' in *Combinatorics* (ed. M. Hall, Jr. and J.H. van Lint), Nato Advanced Study Institute (Series C) 16, Reidel, Dordrecht (1974)
41. O'Nan, Michael E., 'Normal structure of the one-point stabilizer of a doubly-transitive permutation group, I, II', *American Mathematical Society. Transactions*, 214, 1-42, 43-74 (1975)
42. O'Nan, Michael E., 'Doubly transitive groups of odd degree whose one point stabilizers are local', *Journal of Algebra*, 39, 440-482 (1976)
43. Holt, D.F., 'Doubly transitive groups with a solvable one point stabilizer' *Journal of Algebra* [to appear]
44. Neumann, Peter M., 'Transitive permutation groups of prime degree' in *Proceedings of the international conference on the theory of groups, Canberra, 1973* (ed. M.F. Newman), Lecture Notes in Mathematics 372, Springer, Berlin (1974)
45. Neumann, Peter M., *Permutationsgruppen von Primzahlgrad*, Vorlesungen aus dem Mathematischen Institut, Giessen [to appear]
46. Huppert, B., *Endliche Gruppen, I*, Springer, Berlin (1967)
47. Zassenhaus, Hans J., *The theory of groups*, 2nd edn, Chelsea, New York (1958)
48. MacLane, Saunders, *Homology*, Springer, Berlin (1963)
49. Hall, P., 'A note on soluble groups', *London Mathematical Society. Journal*, 3, 98-105 (1928)
50. Carter, R.W., 'Nilpotent self-normalizing subgroups of soluble groups', *Mathematische Zeitschrift*, 75, 136-139 (1961)
51. Hall, P., 'A contribution to the theory of groups of prime power order', *London Mathematical Society. Proceedings*, 36, 29-95 (1933)

52. Lazard, M., 'Sur les groupes nilpotents et les anneaux de Lie', *Ecole Normale Supérieure. Annales Scientifiques,* **71**, 101-190 (1954)
53. Magnus, Wilhelm, Karrass, Abraham and Solitar, Donald, *Combinatorial group theory,* Wiley-Interscience, New York (1966)
54. Hall, Marshall, Jr. and Senior, James K., *The groups of order 2^n ($n \leqslant 6$),* Macmillan, New York (1964)
55. Burnside, William, *Theory of groups of finite order,* 2nd edn, Cambridge University Press, Cambridge (1911)
56. Serre, Jean-Pierre, *Représentations linéaires des groupes finis,* Hermann, Paris (1967)
57. Curtis, Charles R. and Reiner, Irving, *Representation theory of finite groups and associative algebras,* Wiley-Interscience, New York (1962)
58. Dornhoff, Larry, *Group representation theory: Part A, ordinary representation theory: Part B, modular representation theory,* Marcel Dekker, New York (1971, 1972)
59. Feit, Walter, *Characters of finite groups,* Benjamin, New York (1967)
60. Isaacs, I. Martin, *Character theory of finite groups,* Academic Press, New York (1976)
61. Green, J.A., *Vorlesungen über modulare Darstellungstheorie endlicher Gruppen,* Vorlesungen aus dem Mathematischen Institut, Giessen (1974)
62. Hamernik, Wolfgang, *Group algebras of finite groups: defect groups and vertices,* Vorlesungen aus dem Mathematischen Institut, Giessen (1974)
63. Feit, Walter, *Representations of finite groups, I, II,* Lecture Notes, Yale University, New Haven (1969, 1976)
64. Thompson, John G., 'Vertices and sources', *Journal of Algebra,* **6**, 1-6 (1967)
65. Dade, E.C., 'Blocks with cyclic defect groups', *Annals of Mathematics,* **84**, 20-48 (1966)
66. Hall, Marshall, Jr., 'A search for simple groups of order less than one million' in *Computational problems in abstract algebra* (ed. by J. Leech), Pergamon Press, Oxford (1970)
67. Littlewood, Dudley, E., *The theory of group characters and matrix representations of groups,* 2nd edn, Clarendon Press, Oxford (1950)
68. Robinson, G. de B., *Representation theory of the symmetric group,* University of Toronto Press, Toronto (1961)
69. Coleman, A.J., *Induced representations with applications to S_n and GL(n),* Queen's Papers in Pure and Applied Mathematics 4, Queen's University, Kingston, Ontario (1966)
70. Kerber, Adalbert, *Representations of permutation groups, I, II,* Lecture Notes in Mathematics 240, 495, Springer, Berlin (1971, 1975)
71. Green, J.A., 'The characters of the finite general linear groups', *American Mathematical Society. Transactions,* **80**, 402-447 (1955)
72. Carter, Roger W. and Lusztig, George, 'On the modular representations of the general linear and symmetric groups', *Mathematische Zeitschrift,* **136**, 193-242 (1974)
73. Deligne, P. and Lusztig, G., 'Representations of reductive groups over finite fields', *Annals of Mathematics,* **103**, 103-161 (1976)
74. Borel, A. *et al., Seminar on algebraic groups and related finite groups,* Lecture Notes in Mathematics 131, Springer, Berlin (1970)
75. Springer, T.A., 'Caractères de groupes de Chevalley finis' in *Séminaire Bourbaki, 1973,* Lecture Notes in Mathematics 383, Springer, Berlin (1974)
76. McKay, John, *The character tables of the known finite simple groups of order less than 10^6,* Mathematical Institute, Oxford (1970); edited by P.J. Lambert

77. Baumslag, Gilbert (ed.), *Reviews on infinite groups, as printed in Mathematical Reviews, 1940-1970, Volumes 1-40,* 2 vols, American Mathematical Society, Providence, R.I. (1974)

78. Coxeter, H.S.M. and Moser, W.O.J., *Generators and relations for discrete groups,* 3rd edn, Springer, Berlin (1972)

79. Lyndon, R. and Schupp, P.E., *Combinatorial group theory,* Springer, Berlin (1976)

80. Stallings, John, *Group theory and three-dimensional manifolds,* Yale University Press, New Haven (1971)

81. Serre, Jean-Pierre and Bass, Hyman, *Arbres, amalgames et SL_2,* Lecture Notes, Collège de France (1969) [to appear in Lecture Notes in Mathematics, Springer]

82. Miller, Charles, F., III, *On group-theoretic decision problems and their classification,* Annals of Mathematics Studies 68, Princeton University Press, Princeton (1971)

83. Baumslag, Gilbert, 'Finitely presented groups' in *Proceedings of the international conference on the theory of groups, Canberra, 1965* (ed. L.G. Kovács and B.H. Neumann), Gordon and Breach, New York (1967)

84. Baumslag, Gilbert, 'Some problems on one-relator groups' in *Proceedings of the international conference on the theory of groups, Canberra, 1973* (ed. M.F. Newman), Lecture Notes in Mathematics 372, Springer, Berlin (1974)

85. Golod, E.S., 'On nil-algebras and residually finite p-groups', *Akademiya Nauk SSSR. Izvestiya—Seriya matematicheskaya,* **28**, 273-276 (1964); translated in *American Mathematical Society. Translations,* **48**, 103-106 (1965)

86. Alešin, S.V., 'Finite automata and the Burnside problem for periodic groups', *Matematicheskie Zametki,* **11**, 319-328 (1972); translated in *Mathematical Notes,* **11**, 199-203 (1972)

87. Adjan, S.I., *The Burnside problem and identical relations in groups,* Nauka, Moscow (1975); in Russian [a translation is to appear from Springer]

88. Britton, J.L., 'The existence of infinite Burnside groups' in *Word problems* (ed. W.W. Boone, F.B. Cannonito and R.C. Lyndon), North-Holland, Amsterdam (1973)

89. Cohn, P.M., *Universal algebra,* Harper and Row, London (1965)

90. Neumann, Hanna, *Varieties of groups,* Springer, Berlin (1967)

91. Kovács, L.G. and Newman, M.F., 'Hanna Neumann's problems on varieties of groups' in *Proceedings of the international conference on the theory of groups, Canberra, 1973* (ed. M.F. Newman), Lecture Notes in Mathematics 372, Springer, Berlin (1974)

92. Hall, P., *Nilpotent groups,* Queen Mary College, London (1969); notes of lectures given at the Canadian Mathematical Congress, Alberta, 1957

93. Baumslag, Gilbert, *Lecture notes on nilpotent groups,* CBMS Regional Conference Series in Mathematics 2, American Mathematical Society, Providence, R.I. (1971)

94. Hall, P., 'Finiteness conditions for soluble group', *London Mathematical Society. Proceedings,* **4**, 419-436 (1954)

95. Hall, P., 'On the finiteness of certain soluble groups', *London Mathematical Society. Proceedings,* **9**, 595-622 (1959)

96. Hall, P., 'The Frattini subgroups of finitely generated groups', *London Mathematical Society. Proceedings,* **11**, 327-352 (1961)

97. Robinson, Derek J.S., *Finiteness conditions and generalised soluble groups,* 2 vols, Springer, Berlin (1972)

98. Robinson, Derek J.S., 'A new treatment of soluble groups with finiteness conditions on their Abelian subgroups', *London Mathematical Society. Bulletin,* **8,** 113-129 (1976)
99. Kegel, Otto H. and Wehrfritz, Bertram A.F., *Locally finite groups,* North-Holland, Amsterdam (1973)
100. Plotkin, B.I., *Groups of automorphisms of algebraic systems,* Nauka, Moscow (1966); translation published by Wolters-Noordhoff, Groningen (1972)
101. Dixon, John D., *The structure of linear groups,* Van Nostrand Reinhold, London
102. Wehrfritz, B.A.F., *Infinite linear groups,* Springer, Berlin (1973)
103. Passman, Donald S., *Infinite group rings,* Marcel Dekker, New York (1971)
104. Passman, D.S., 'What is a group ring?', *American Mathematical Monthly,* **83,** 175-185 (1976)
105. Cartan, Henri and Eilenberg, Samuel, *Homological algebra,* Princeton University Press, Princeton (1956)
106. Hilton, P.J. and Stammbach, U., *A course in homological algebra,* Springer, Berlin (1971)
107. Babakhanian, Ararat, *Cohomological methods in group theory,* Marcel Dekker, New York (1972)
108. Atiyah, M.F. and Wall, C.T.C., 'Cohomology of groups' in *Algebraic number theory* (ed. J.W.S. Cassels and A. Fröhlich), Academic Press, London (1967)
109. Gruenberg, Karl W., *Cohomological topics in group theory,* Lecture Notes in Mathematics 143, Springer, Berlin (1970)
110. Stammbach, Urs, *Homology in group theory,* Lecture Notes in Mathematics 359, Springer, Berlin (1973)
111. Leedham-Green, C.R. *et al.,* 'Homology in varieties of groups, I-V', *American Mathematical Society. Transactions,* **162,** 1-33 (1971), **170,** 293-303 (1972); *Acta Mathematica* [to appear]
112. Kaplansky, Irving, *Infinite Abelian groups,* rev. edn, University of Michigan Press, Ann Arbor, Mich. (1969)
113. Fuchs, Laszló, *Infinite Abelian groups,* 2 vols, Academic Press, New York (1970, 1973)
114. Shelagh, Saharon, 'Infinite Abelian groups, Whitehead problem and some constructions' and 'A compactness theorem for singular cardinals, free algebras, Whitehead problem and transversals', *Israel Journal of Mathematics,* **18,** 243-256 (1974); **21,** 319-349 (1975)
115. Borel, Armand, *Linear algebraic groups,* Benjamin, New York (1969)
116. Humphries, James E., *Linear algebraic groups,* Springer, Berlin (1975)
117. Bourbaki, N., *Éléments de mathématiques: Groupes et algèbres de Lie,* Hermann, Paris: Fasc. XXVI, Chapitre I (1960); Fasc. XXXVII, Chapitres II, III (1972); Fasc. XXXIV, Chapitres IV, V, VI (1968); Fasc. XXXVIII, Chapitres VII, VIII (1975)
118. Chevalley, Claude, *Theory of Lie groups,* Princeton University Press, Princeton (1946)
119. Tits, Jacques, *Buildings of spherical type and finite BN-pairs,* Lecture Notes in Mathematics 386, Springer, Berlin (1974)
120. Weyl, Hermann, *The classical groups: their invariants and representations,* 2nd edn, Princeton University Press, Princeton (1946)
121. Boerner, H., *Darstellungen von Gruppen,* 2nd edn, Springer, Berlin (1967)
122. Adams, J. Frank, *Lectures on Lie groups,* Benjamin, New York (1969)
123. Bruck, Richard Hubert, *A survey of binary systems,* Springer, Berlin (1958)
124. Clifford, A.H. and Preston, G.B., *The algebraic theory of semigroups,* 2 vols, American Mathematical Society, Providence, R.I. (1961, 1967)

125. Rédei, László, *The theory of finitely generated commutative semigroups*, Pergamon Press, Oxford (1965)
126. Howie, J.M., *An introduction to semigroup theory*, London Mathematical Society Monographs 7, Academic Press, London (1976)
127. Arbib, Michael A., *Theories of abstract automata*, Prentice-Hall, Englewood Cliffs, N.J. (1969)
128. Eilenberg, Samuel, *Automata, languages and machines*, 2 vols, Academic Press, New York (1973, 1976)
129. Wells, Charles, 'Some applications of the wreath product construction', *American Mathematical Monthly*, **83**, 317-338 (1976)
130. Rhodes, John *et al.*, 'Global structure theories for finite semigroups', *Advances in Mathematics*, **11**, 157-266 (1973) [introduction and four articles]

10

Measure and Probability

S.J. Taylor

10.1 THEORY OF MEASURE

This is one of the basic tools of twentieth century analysis. A good understanding of measure is essential if one is to study recent developments in almost all areas of analysis or even its application to mathematical physics. We start by describing the main ideas. For most people, a logical account of the structure of measure theory in its fully developed form is not the best source for a first encounter. It is better to start by understanding some of the important problems of analysis in the era around 1900— for those were the driving force which gave birth to a new subject.

When we have our first introduction to the calculus we quickly learn that, if

$$F(x) = \int_a^x f(t)\, dt$$

then the derivative $F'(x)$ exists and equals $f(x)$ at least at points x where f is smooth. This means that we can consider differentiation and integration as inverse processes — and this is the means used for evaluating standard integrals. In 1881 Volterra, who was a student of Dini in Pisa, published an example of a function F, defined on the interval $(0, 1)$, whose derivative $f = F'$ exists everywhere, is bounded, but is not integrable in the sense of Riemann — who had given the only rigorous definition of integrability available at that time. This example is based on the function $g(x) = x^2 \sin 1/x$ for $x \neq 0$, $g(0) = 0$, which has a discontinuous derivative at $x = 0$; Volterra patches together lots of such functions to spread the discontinuities to all points in a set E of positive length. Henri Lebesgue[1], in his thesis of 1902, noted Volterra's example, and accepted as one of his objectives the definition of an integral which would make differentiation and integration inverse operators for a much wider class of functions. Another clear need at this time was for a definition of the integral which would allow term-by-term integration of a convergent series.

The crucial step needed in modifying the Riemann integral was the development of 'length' for a linear set. By the end of the nineteenth century, the theory of cardinals had been sufficiently developed to clarify the meaning of a countable set. A study of the topology of R had led to the notion of completeness and the theorem of Baire that a complete metric space cannot be expressed as a countable union of closed nowhere dense sets. Emile Borel first extended the notion of length from intervals in R to a wide class of sets. Lebesgue discovered the associated integration theory, and in doing so found it necessary to extend further Borel's notion of length, to what we now call Lebesgue measure on R. Time spent in absorbing the flavour of real analysis around 1900 will provide a clear motivation for a study of measure. The original works of Lebesgue[1-3] are well worth reading, as are the historical notes by Bourbaki[4] in the introduction to his book on integration, and there is a wealth of fascinating material in Saks[5].

The contribution of E. Borel is assessed by Fréchet[6] (pp. 53-63). His main idea was the importance of countable additivity: in other words, if E_1, E_2, . . ., E_n, . . ., is a sequence of disjoint sets for which the measure is defined, and

$$E = \bigcup_{i=1}^{\infty} E_i,$$

then the measure of E should be defined, and

$$mE = \sum_{i=1}^{\infty} mE_i \tag{10.1}$$

Peano has previously defined inner and outer content for planar sets and had shown that a bounded function f is Riemann integrable over $[a, b]$ if and only if these two contents are the same for the 'area under the graph'. Jordan developed this to obtain a theory of finitely additive set functions, but Borel insisted that the appropriate class $\mathscr{A}$ of subsets of X on which to define m must be a σ-field (or σ-algebra, or Borel field) or class closed under complementation and countable unions (and therefore also countable intersections). A set function $m: \mathscr{A} \to R^+$

which assigns to each set $E \in \mathscr{A}$ a non-negative real number (or $+\infty$) is called a *measure* if it satisfies (10.1), and $(X, \mathscr{A}, m)$ is then called a measure space. Even for $X = \mathsf{R}$, it is not immediately obvious that a non-trivial measure exists. Lebesgue constructed a σ-algebra $\mathscr{M}$ of subsets of R which contains open intervals (and therefore all Borel sets) and a measure m on $\mathscr{M}$ such that; when

$$I = (a, b), \; mI = b - a \tag{10.2}$$

for any

$$E \in \mathscr{M}, \; m(x + E) = mE \tag{10.3}$$

where $x + E$ is the result of adding the fixed real x to the elements of E (this property, 10.3, is called translation invariance).

Even though it is of peripheral importance in the subsequent development, it is worth asking whether or not it is possible to extend the definition of m to all subsets of R while preserving Equations (10.1)–(10.3). Using the axiom of choice, Vitali proved in 1905 that the answer is negative, but that result opened up the so-called 'problem of measure'. In R^k, is it possible to define a non-trivial *finitely* additive set function μ on all subsets such that μ is finite for bounded sets and equal for congruent sets? An application of the Hahn–Banach theorem (which in turn assumes the axiom of choice) shows that, in R and R^2 it is possible to define such a μ which coincides with Lebesgue measure m on $\mathscr{M}$. That the opposite is true in R^3 (and therefore in R^k, $k \geqslant 3$) follows from the Banach–Tarski paradox, which can be formulated as follows. Let S, T be disjoint solid spheres with the same radius. There exist disjoint sets $E_1, E_2, \ldots, E_{41} \subset S; F_1, F_2, \ldots, F_{41} \subset S \cup T$ such that E_i is congruent to F_i for each i and

$$S = \bigcup_{i=1}^{41} E_i, \; S \cup T = \bigcup_{i=1}^{41} F_i$$

If μ were a solution to the problem of measure, then $\mu S = \mu T$, $\mu E_i = \mu F_i$, so that

$$\mu S = \Sigma \mu E_i = \Sigma \mu F_i = \mu S \cup T = \mu S + \mu T = 2\mu S$$

and μ would be zero for every sphere and therefore for every bounded set. There is a good account of these problems in Sierpinski[7].

The details of the construction of Lebesgue measure are deep, and it is now clear from subsequent generalisations that special measures on R are no easier to obtain than a general measure space $(X, \mathscr{A}, m)$. It is therefore more efficient to learn a definition of Lebesgue measure which extends to a general $(X, \mathscr{A}, m)$ or to learn about general measure spaces directly and treat Lebesgue measure as an important special case which provides motivation and clarification of the meaning of the results. There are many accounts of measure theory which use the process of extension devised by Carathéodory in 1914. Perhaps Munroe[8] gives the clearest explanation of the use of a 'pre-measure' to obtain an outer measure defined on all subsets (outer measures are sub-additive). This outer measure becomes a measure when restricted to the σ-algebra $\mathscr{M}$ of 'measurable' sets. This method automatically yields a complete measure space $(X, \mathscr{M}, \mu)$; that is, one for which $A \in \mathscr{M}$, $\mu A = 0$, $B \subset A$ implies that $B \in \mathscr{M}$ (and therefore $\mu B = 0$). The book by Taylor[9] uses a variant of this approach in which the starting point is a measure defined on a very restricted class $\mathscr{C}$ of sets, and the extension given by Carathéodory is defined on the completion of the σ-algebra generated by $\mathscr{C}$.

Given a measure space $(X, \mathscr{M}, \mu)$ an extended real-valued function $f : X \to \mathsf{R}^*$ is said to be measurable if, for every open set $G \subset \mathsf{R}^*$, $f^{-1}(G) \in \mathscr{M}$. Thus measurability can be thought of as smoothness relative to $\mathscr{M}$. If X has a topology, it is usual for $\mathscr{M}$ to contain the open sets (and therefore the Borel sets): in this case measurability is clearly a weaker condition than continuity. The measure μ is said to be regular if, for every $\epsilon > 0$, $E \in \mathscr{M}$, there is a closed F and open G with $F \subset E \subset G$ such that $\mu(G \backslash F) < \epsilon$. One of the key theorems on measurable functions is usually called Lusin's theorem, although the first proof is due to Vitali in 1905. It says that if μ is a regular measure on a complete measure space $(X, \mathscr{M}, \mu)$ and X is locally compact, then, for each $\epsilon > 0$ and measurable $f : X \to \mathsf{R}$, which vanishes outside a set $E \in \mathscr{M}$ with $\mu E < \infty$, there is a continuous function $g : X \to \mathsf{R}$ such that $\mu\{x : f(x) \neq g(x)\} < \epsilon$. So measurable functions are the same as continuous functions outside a set of small measure. A good account of these results is to be found in Natanson[10].

Given a measure space $(X, \mathscr{M}, \mu)$ and a measurable function $f : X \to \mathsf{R}^*$, there are several different ways of defining the absolute integral $\int f \mathrm{d}\mu$. Lebesgue's idea for integrals on R was to replace Riemann's dissection of the domain by dissections of the range to give approximating sums

$$S_\alpha = \Sigma y_{i+1} mA_i; \quad s_\alpha = \Sigma y_i mA_i$$

where $A_i = \{x : y_i \leqslant f(x) < y_{i+1}\}$. Thus measure theory is used to collect together 'approximately equal' values of $f(x)$ in a delicate manner. A modern account of this method of defining the integral is given by Benedetto[11] — its disadvantage is that it does not give directly the integral over a set of infinite measure — and extension to this case is not even possible when $(X, \mathscr{M}, \mu)$ is such that X cannot be expressed as a countable union of sets of finite measure. A quicker and more general method is to define the integral first for non-negative simple functions – which take constant values on each set of a partition of X, and then extend the definition by monotone convergence to non-negative measurable f. Since any measurable function is the difference of two non-negative ones, a measurable function is integrable if both pieces give an integral with finite answer. This method of obtaining the integral can be found in Reference 9.

Once the integral has been defined, there is a sequence of theorems which follow naturally, although the order depends on the particular treatment being followed. Any standard textbook on 'Measure and Integration' gives these results – but Royden[12] gives a well-motivated account of the properties of the integral. Perhaps the most used property of all is the 'majorised convergence theorem', which says that if f, g and $\{f_n, n = 1, 2, \ldots\}$ are measurable functions, g is integrable and

$$f_n(x) \to f(x), \quad |f_n(x)| \leqslant g(x) \text{ for all } x$$

then

$$\int f_n \, d\mu \to \int f \, d\mu$$

We saw that one of the problems which led Lebesgue to define his integral was the need to extend the class of functions for which integration and differentiation are inverse processes. For any Lebesgue integrable function f, the indefinite integral

$$F(x) = \int\limits_{-\infty}^{x} f(t) \, dt$$

is absolutely continuous: that is, $\Sigma |F(b_i) - F(a_i)|$ is small for any finite collection of disjoint intervals (a_i, b_i) of small total length. Further, $F'(x) = f(x)$ for all x except a set of zero measure, and it is possible to give a descriptive definition of the Lebesgue integral by restricting the class of primitive functions to be absolutely continuous. Although the Lebesgue integral solves the original problem of Volterra (where the derivative $F'(x)$ was bounded) it is not general enough to integrate the derivative $f(x) = F'(x)$ of every function with a finite derivative for all x. Two different, but equivalent, methods of extending the definition of an integral on R, one due to Perron, the other due to Denjoy, were developed to give a wide class of functions F which are the integral of their derivative. For each integral, a descriptive definition has also been devised using the properties which characterise the primitive functions F which can be obtained. A clear account of this branch of integration theory is given by Saks[5].

Lebesgue's notion of absolute continuity for real functions generalises to a measure space $(X, \mathcal{M}, \mu)$. A σ-additive set function τ on $\mathcal{M}$ is said to be absolutely continuous (with respect to μ) if, given $\epsilon > 0$, there is a $\delta > 0$ such that

$$E \in \mathcal{M}, \ \mu E < \delta \Rightarrow |\tau E| < \epsilon$$

It is clear that, for any integrable f,

$$\tau E = \int_E f(x)\mu(\mathrm{d}x) \qquad (10.4)$$

defines an absolutely continuous τ. The converse of this result is the celebrated Radon–Nikodym theorem. Given a τ which is absolutely continuous, there is an f such that τ is given by Equation 10.4. Under suitable additional conditions (certainly satisfied on R), f can be obtained as a pointwise derivative of τ with respect to μ. The best reference for results up to 1930 is again Reference 5 but modern extensions, in particular to the case of finitely additive τ, are given in Dunford and Schwartz[13].

For many purposes we have to identify two measurable functions f_1, f_2 such that $\{x : f_1(x) \neq f_2(x)\}$ has zero measure. The abstract method is to take the equivalence class of functions equal a.e. to f as the new object of study. It is only in this sense that the Radon–Nikodym derivative is unique. However, it is obviously attractive to have a procedure which picks a canonical representative of each equivalence class. If one just uses the axiom of choice to bring this about, one loses all smoothness from the situation. The problem of lifting concerns the definition of a choice function which preserves the algebraic structure. In its strong form one wants a choice function ρ such that

      (i) $\rho(f) \equiv f$
      (ii) $f \equiv g \Rightarrow \rho(f) = \rho(g)$
      (iii) $\rho(1) = 1$
      (iv) $f \geqslant 0 \Rightarrow \rho(f) \geqslant 0$
      (v) $\rho(af + bg) = a\rho(f) + b\rho(g)$
      (vi) $\rho(fg) = \rho(f)\rho(g)$

where the congruence is that of equality a.e. A clear account of our present knowledge about the existence of liftings, and some of their uses, is given by the Ionescu-Tulceas[14].

The use of measure and absolute integration has led to a rapid expansion of many areas of analysis largely because very few conditions (and only natural ones) are needed or applying the tool. For example, the theory of trigonometric series was explored in the nineteenth century by Riemann and others; it becomes much simpler using the Lebesgue integral. Given any complete orthonormal sequence $\{f_n\}$ of measurable functions such that f_n^2 is integrable, there is a unique expansion

$$f(x) = \sum_{n=1}^{\infty} c_n f_n(x)$$

where equality is to be understood in the sense that

$$\int \left(f - \sum_{r=1}^{n} c_r f_r \right)^2 \mathrm{d}\mu \to 0$$

A very full and readable account of the theory of trigonometric series is given by Zygmund[15]. It is worth noting that this is the start of a new subject, called harmonic analysis. We can think of trigonometric functions as being defined on the circle group $T = [0, 2\pi]$ in which addition is carried out modulo 2π. This is an example of a topological group, since the group operation is continuous with respect to the usual topology. Given any compact topological Abelian group G, there is a unique measure μ on G which is invariant under the operation of translation by an element of G (this is the property, 10.3, of Lebesgue measure). This measure is called Haar measure on G. The definition extends to the non-Abelian case and to the σ-compact case. Harmonic analysis is the study of functions and measures defined on topological groups. Fourier analysis can be done on any locally compact Abelian group. For a beginning study of harmonic analysis see Katznelson[16]; a more general account is given in Hewitt and Ross[17].

The study of ergodic theory is a mathematical abstraction arising out of the Gibb's theory of statistical mechanics. When μ is Haar measure on a group G, then μ is invariant under the group operation $E \to a * E$, for $a \in G$. In a general measure space $(X, \mathcal{M}, \mu)$, measure-preserving transformations $T : X \to X$ are those for which $\mu(TE) = \mu E$ for all $E \in \mathcal{M}$. In this case it is clear that the iterates T^k of T are also measure-preserving. If f is integrable, then

$$\frac{1}{n} \sum_{i=0}^{n-1} f(T^i x)$$

converges for almost all x to an integrable f^* which is invariant under T (that is, $f^*(Tx) = f^*(x)$ almost everywhere). If $\mu X < \infty$, then $\int f^* \mathrm{d}\mu = \int f \mathrm{d}\mu$. The transformation T is said to be ergodic (a kind of mixing condition) if there are no proper invariant subsets of X. In this case the limit function f^* above has to be constant. This means that the space mean becomes equal almost everywhere to the time mean

$$\lim_{n \to \infty} \frac{1}{n} \Sigma f(T^i x).$$

An introduction to this fascinating and important area of mathematics can be found in Halmos[18] or, for an approach which illustrates the theory by intuitive geometric arguments in $[0, 1]$, see Friedman[19].

The relationship between measure theory and functional analysis is of the utmost importance. In fact, one popular way of defining measure is to use the techniques of functional analysis. If X is locally compact, $C_0(X)$ denotes the Banach space of continuous $f : X \to \mathsf{R}$ with the property that for every $\epsilon > 0$, there is a compact K such that $|f(x)| < \epsilon$ for all x outside K. The norm is

$$\|f\|_\infty = \sup_{x \in X} |f(x)|$$

$M(X)$, the space of bounded regular signed measures on $(X, \mathscr{B})$, where $\mathscr{B}$ is the σ-field of Borel sets, is also a Banach space with norm

$$\|\mu\| = |\mu|(X)$$

The Riesz representation theorem states that the dual space $C_0(X)'$ of linear functionals on $C_0(X)$ is isometrically isomorphic to $M(X)$; that is, for each $L \in C_0(X)'$ there is a unique $\mu \in M(X)$ such that

$$L(f) = \int f \mathrm{d}\mu \text{ for all } f \in C_0(X)$$

This theorem has a serious defect which reduces its usefulness, namely that $M(X)$ is too small a set: in fact if $X = \mathsf{R}$ with the usual topology, $M(\mathsf{R})$ does not even contain Lebesgue measure m, since $m\,\mathsf{R} = \infty$. In order to get a larger class of functionals, we need to take a smaller Banach space. The set $C_K(X)$ is the class of continuous $f : X \to \mathsf{R}$ for which the support of f, the closure of the set $\{x : f(x) \neq 0\}$ is compact. The space $C_K(X)'$ of linear functionals $\mu : C_K(X) \to \mathsf{R}$ is the space of Radon measures on X. When a Radon measure is restricted to a compact subset of X, it becomes a linear functional on $C_0(X)$. On the whole space X it is σ-finite, but need not be finite. A good place to learn about these general representation theorems is in Hewitt and Stromberg[20].

Many authors have used the connections observed in the last paragraph as a device for obtaining measures out of a study of linear functionals. A good exposition of this reverse order is given by Riesz and Sz.-Nagy[21], but it is also used by Bourbaki[4] and many other authors. They start with $C_K(X)$ and obtain properties of the dual space $C_K(X)'$. Although the elements of this dual space are not initially defined as a set function, all the classical properties of a σ-additive set function can be deduced. The positive linear functionals in $C_K(X)'$ can be identified as positive measures on the σ-algebra $\mathscr{B}$ of Borel sets which assign finite measure to every compact set. Any reader who feels at home in functional analysis can learn measure theory via this route:

for the majority of analysts, the historical route I have outlined in this chapter is intuitively easier to follow.

Another disadvantage of the approach to measure via linear functionals is that although the majority of measures in everyday use are Radon measures, there are useful and interesting measures which cannot be obtained in this way. We may be interested in an abstract measure space with no topology, or the wrong topology. Or we may have a good topology and want a measure which is not finite on compact sets. In Euclidean space R^k (the method generalises to a metric space) Hausdorff defined an important class of such measures which are not Radon measures, although they share with Lebesgue measure the property of invariance under isometric transformations. One starts with a real function $\phi : R^+ \to R^+$ which is monotone and satisfies

$$\lim_{s \downarrow 0} \phi(s) = 0.$$

For any subset $E \subset R^k$, define an outer measure by

$$\phi - m^*E = \lim_{\substack{\delta \downarrow 0}} \inf_{\substack{E \subset \cup C_i \\ d(C_i) < \delta}} \Sigma \phi(d(C_i)) \tag{10.5}$$

where $d(C_i)$ denotes the diameter of the set C_i, and the infimum in Equation 10.5 is taken over countable covers of E by sets C_i, with diameter $< \delta$. This outer measure determines a class $\mathscr{M}_\phi$ of ϕ-measurable subsets, and a measure on $\mathscr{M}_\phi$. If $\phi(s)/s^k \to \infty$ as $s \downarrow 0$, it is easy to see that ϕ is not σ-finite on a closed ball. A recent account of the theory of Hausdorff measures can be found in Rogers[22]. These measures have very significant geometrical properties when $\phi(s) = s^m$ and it is an integer. They are the starting point for the subject of geometric measure theory. The best book to survey this area is by Rado[23].

Measure theory is also used to obtain additional information about sets which are known to be small. For example, in the L_2-theory of trigonometric series, convergence pointwise occurs except for an exceptional set of zero Lebesgue measure. Different measures, such as those of Hausdorff mentioned above, give information about the 'size' of the exceptional set. There are many problems of this type in classical harmonic analysis, and some of them extend to the abstract situation. An interesting recent account of this area is given in Lindahl and Poulsen[24].

It is worth pointing out that there is yet another distinct way of introducing measure. Instead of using the whole apparatus of functional analysis, Daniel extracted some specific techniques and obtained a relatively simple definition of an integral. He starts with a relatively simple class of functions – the step functions on R – where $\int f$ has an obvious value – and extends the class by taking monotone limits. The resulting integral is equivalent to the Lebesgue integral on the line, so that Lebesgue measure can be obtained as $mE = \int I_E$, where I_E denotes the characteristic function of E, and all the usual properties can be extracted. A careful simplified version of this method is used in the book by Weir[25].

The richness of mathematics results from the interplay of ideas and techniques from apparently unrelated branches. If we were to try to survey all the areas where measure theory is used, the whole of this volume would be needed. For example Dieudonné[26] first defines the integral and obtains measure, but he then explores the connections with Lie groups, Riemannian geometry, differential topology and multilinear algebra; Dinculeanu[27] studies measures which take values in a Banach space rather than in R; and Fremlin[28] uses the lattice structure to define measure on a Boolean ring. We will now describe in detail one area which owes its existence in modern form to its basis in measure, and to the large analytical apparatus which can now be easily developed.

10.2 PROBABILITY THEORY

Analysis of gambling games provided the early stimulus to probability. The basic idea was that of fairness, or symmetry, which means that each of the possible results of the game must be equally likely. (Thus, if you toss a coin it shows head or tail each with probability 1 in 2; if you throw a cubical die, each face shows up with probability 1 in 6; if you roll a ball in roulette it comes to rest in a given slot with probability 1 in 37; a well-shuffled pack of playing cards has a given order with probability 1 in 52! etc.). There are at least two drawbacks to this idea as a basis for probability theory. The first is that it cannot deal with the situation where the number of possible results is infinite; and the second is that 'life is unfair' – so that, in the real world, even when the number of possible outcomes is finite, there is usually a manifest lack of symmetry, so that the model of 'equally likely outcomes' is quite unsuitable.

Early attempts to generalise the model of equally likely events to the infinite case used calculus and led to a study of 'geometric probability'. It was quickly discovered that questions such as 'What is the probability that a chord of a circle is longer than the radius?' do not have a unique answer: in fact at least three methods of calculation can be justified as plausible and lead to three different answers. This is called Bertrand's paradox.

The situation at the beginning of the twentieth century was clearly unsatisfactory, and most serious books written at this time contained a lot of philosophy — attempting to define actual objects rather than relations. At this stage statistics became an object of study, and this quickly forced a realisation that the classical ideas of probability were inadequate. Von Mises introduced the idea of the collective — which we now call the sample space of all conceivable outcomes of an experiment — and defined the probability of an event as the limit of the frequency ratio of the number of times the event happens in n trials, divided by n. An axiomatic approach in which probabilities are defined in terms of limiting frequencies has not been widely adopted because of its complexity. But von Mises' notion of a sample space is the key to the axiomatic basis, clearly formulated by Kolmogorov[29], which is widely accepted today. In this model we consider a measure space $(\Omega, \mathscr{F}, P)$ in which the measure P satisfies $P(\Omega) = 1$, and events E are subsets in $\mathscr{F}$ which are said to occur with probability $P(E)$. This is now universally accepted as the right basis for probability theory, and few probabilists today give any thought to the foundations. There are still statisticians who worry about such matters, and a reader interested in these philosophical questions will get a recent analysis in Gillies[30].

Perhaps the reader is now tempted to conclude that probability is just a special case of finite measure theory, with the extra normalising condition that $P(\Omega) = 1$. In a sense this is true, but the additional structure imposed by statistical concepts brings with it a rich theory which becomes visible in quite simple situations. Therefore I would advise the reader that, before he attempts to master abstract probability theory in its full generality, he should first read Feller[31], as this is the book which, more than any other, has brought probability into the normal curriculum of an undergraduate mathematics degree. One reason for the success of Reference 31 is the brilliant collection of diverse and vivid examples and applications with which the theory is illustrated. There are two sides to the development of probability theory: one is rigorous axiomatic work using the tools of measure theory; the other is the development of thought patterns using gambling situations, coin tossing, motions of a physical particle. Both are important in establishing an intuitive feeling for the subject. The strength of Reference 31 is in helping to establish the right thought patterns. This book considers only a discrete sample space Ω, with a countable number of points $\{\omega_i\}$. To each ω_i is assigned the probability $p_i, 0 \leqslant p_i \leqslant 1$, where

$$\sum_{\text{all pts}} p_i = 1.$$

This means that $\mathscr{F}$ can be the class of all subsets of Ω with P defined by

$$PE = \sum_{\omega_i \in E} p_i$$

Integration is reduced to summing an absolutely convergent series, and the whole theory can be developed without the apparatus of measure theory — although it turns out that all the techniques can be thought of as special cases of measure theoretic theorems.

Perhaps the most important single concept in probability theory is that of independence. This is a mathematical abstraction of the notion that knowledge than an event A has occurred does not affect the probability of a second event B. We say that A and B are independent o if

$$P(A \cap B) = PA \cdot PB$$

This extends to a collection of events $\{A_\alpha\}$, $\alpha \in I$. These are said to be independent if PA_α is not affected by the knowledge of any finite number of other events in the collection. The mathematical tool which allows one to generate independent events is the notion of product measures on the Cartesian product of measure spaces — and a sequence of independent events can be obtained by considering product measure on a countable product. Consider a sequence of Bernoulli trials, in which one observes the number of times an event E occurs in n independent trials and computes the relative frequency $r_n(E)$. By looking at the n-fold product measure P_n it is easy to show that, for every $\epsilon > 0$,

$$P_n\{|r_n(E) - p| > \epsilon\} \to 0 \quad \text{as} \quad n \to \infty$$

where $p = PE$ is the probability of the event E. This is called the 'weak law of large numbers': it makes precise the idea that the relative frequency of occurrence of E is 'likely' to be close to p when the number of trials is large. It is not possible to deduce from this the strong law of large numbers which can be stated

$$P_\infty\{\lim r_n(E) = p\} = 1 \tag{10.6}$$

where now P_∞ denotes the product measure in the countable Cartesian product of copies of $(\Omega, \mathscr{F}, P)$. This provides the link with the von Mises axioms: by taking a measure space as the basic object, it is possible to *prove* that, for almost all sample sequences, $r_n(E)$ has the limiting behaviour which von Mises postulated. The laws of large numbers are valid in more general situations than a sequence of Bernoulli trials, and the proof of the strong law in the general case is quite deep. There is a sense in which the ergodic theorem is equivalent to the strong law. Almost any basic text in probability theory includes a discussion of the laws of large numbers. For example, there is a clear account in Chapter 2 of Lamperti[32] or Chapter 13 of Kingman and Taylor[33]. Another key concept is that of a random variable. The name was given because it was thought of as a numerical measurement of a random phenomenon. However the name is unfortunate because a random variable is *not* a variable, nor is it random! It is now precisely defined as a measurable function $X : \Omega \to$ R on a probability space $(\Omega, \mathscr{F}, P)$. The study of the properties of random variables forms a major part of probability theory. Two important parameters are the average, mean, or expectation, defined by

$$\mu = E(X) = \int X(\omega)P(d\omega)$$

when this exists as an absolute integral; and the variance $\sigma^2(X) = E(X - \mu)^2$. An example of a useful inequality is that due to Chebyshev:

$$P\{|X-\mu| > \lambda\sigma\} \leqslant \frac{1}{\lambda^2} \cdot$$

With the measure theory model for probability theory, this type of result becomes a trivial application of integration theory.

When a probabilistic thinks of a random variable, he does not usually go back to the basic definition of a function $X : \Omega \to$ R on a measure space, because his interest is in the values $X(\omega)$ and how they are distributed. There are several equivalent ways of following up this interest: the most basic is by study of the distribution function, defined by

$$F(x) = P\{\omega : X(\omega) \leqslant x\}$$

Clearly, $F :$ R $\to [0, 1]$ is monotone increasing, continuous on the right with left limits everywhere, and

$$\lim_{x \to +\infty} F(x) = 1, \quad \lim_{x \to -\infty} F(x) = 0.$$

$\mathscr{D}$ denotes the set of such functions F. To every $F \in \mathscr{D}$ there corresponds a Lebesgue–Stieltjes measure μ_F on R which satisfies, for the probability space (R $, \mathscr{B}, \mu_F$),

$$\mu_F\{t \in \text{ R}; t \leqslant x\} = F(x)$$

This means that the random variable $X(t) = t$, defined on this probability space has distribution function F. Theorems on mappings between measure spaces then allow us to find all the properties of the random variable X in terms of this particular realisation as a measure μ_F on R .

For real t we can always define

$$\phi_X(t) = E \exp(itX)$$

for a given random variable X. The function ϕ_X is called the characteristic function of X. Since there is a $(1, 1)$ correspondence between distribution functions and characteristic functions, the latter are an alternative tool for studying the behaviour of random variables. We say that a family $\{X_\alpha\}$ of random variables is independent if knowledge of the values taken by any finite number of them has no affect on the distribution of the remainder. The power of characteristic functions as a tool comes from the fact that, if X and Y are independent, and $Z = X + Y$, then

$$\phi_Z = \phi_X \phi_Y$$

The study of characteristic functions is an important special case of Fourier analysis on R. The book by Bochner[34] is a key reference for this study, but adequate introductions can be found in Chapter 8 of Breiman[35] or Chapter 12 of Reference 33.

One problem of great importance in the historical development is the question: 'Which probability distributions can arise as the limit of a sequence of distributions each corresponding to a finite sum of independent random variables?' Bernoulli and de Moivre in the eighteenth century knew that if the random variables being added have finite variance, and the sum is suitably standardised, then the limit distribution is normal. This result is called the central limit theorem: it is the basis for the theory of errors, relevant to many experimental situations. Lindeberg and Lévy obtained sufficient and necessary conditions for the limit distribution to be normal. Paul Lévy was responsible for broadening the classical central limit problem by omitting the condition of finite variance for the summands, and a study of his account of these problems[36] gives good insight into the situation. A more systematic account of the results is given in Gnedenko and Kolmogorov[37], and there is a valuable summary in Chapter 6 of Tucker[38].

The limit laws we discussed in the last paragraph are weak laws — they concern only the distribution of the sum. A detailed study of the problem of convergence of measures is given in Billingsley[39]. Another interesting question concerns the asymptotic behaviour of the actual sequence of sums of random variables. As a special case of this we can ask: 'In the strong law of large numbers (10.6), how fast does $r_n(E)$ converge to p?' The answer to this is the celebrated law of iterated logarithm, which we can state as

$$P\left\{ \limsup_{n \to \infty} (r_n(E) - p)\sqrt{\frac{n}{\log\log n}} = c \right\} = 1$$

where c is a finite positive constant depending on p. This kind of asymptotic behaviour occurs when the central limit theorem holds; other types of behaviour are possible for other limit distributions. A systematic account of the older parts of this theory is given in Reference 37; there is a proof of the law of iterated logarithm, and a discussion of the Borel-Cantelli lemmas in Chapter 13 of Reference 33.

10.3 STOCHASTIC PROCESSES

In real life random variables rarely occur singly: they arise in families, and very often there is a natural parameter, time, by which they can be indexed. It is reasonable to ask for a mathematical theory which describes the random evolution of some measurement. The theory of stochastic processes has grown from this stimulus; it has mushroomed in the last 25 years, not only because of its intrinsic interest, but because of its close connections with many diverse areas of mathematics. it feeds on analytic techniques from measure theory, potential theory, Fourier analysis, semi-groups of operators in a Banach space, spectral theory and ergodic theory; and in turn it has applications to topology, functional inequalities, differential equations, information theory, and through the stochastic integral to several areas of mathematical physics. Thus 'stochastic processes' is a good modern example of an area of mathematics which has been stimulated by its applications, while itself leading to extensive research in more established areas in order to develop the techniques needed. The grandfather of all stochastic processes in continuous time is Brownian motion – in the sense that it was the first to be rigorously defined and studied in depth. Paul Lévy uncovered a series of surprising results. His point of view was that of a particle sitting at $X(t)$ and travelling along the path as t grows. His own account[40] is well worth study, although we should remark that he always assumed that future development of X is independent of the past, even if t, the time epoch, is random – in modern terminology, he happily uses the strong Markov property even though this had been neither formulated nor proved at the time. A much more extensive account of Brownian motion and the related diffusion processes is given by Itô and McKean[41]; many books contain good introductions, including Breiman[35], but perhaps the most readable extensive account is given by Freedman[42].

The general theory of stochastic processes was given a mathematical framework by Kolmogorov[29], who used the Carathéodory method of extending measures to set up measure in function space. However, Doob was responsible for illuminating the technical details and setting up a rigorous framework which has been used ever since. His account[43] is a classic which repays careful study; but see Karlin[44] for a less technical account motivated by applications. The first difficulty is that the natural σ-algebra does not contain enough sets, and a simple set like $\{a \leqslant X(t) \leqslant b$ for all t in an interval I$\}$ would be non-measurable, and therefore not have a probability assigned to it. This problem is cured by the condition of separability, which we can think of as ensuring that the values of $X(t)$ are all determined when they are known for t in a suitable countable dense set. This leads on to the establishment of measurability properties for $X(t, \omega)$ in the product space $T \times \Omega$, which are needed to justify the use of integration methods. Very mild additional regularity conditions allow one to assume that, for fixed ω, $X(t, \omega)$ has only simple discontinuities: in this case it is usual to assume that each sample path is right continuous with left limits everywhere.

An important tool in the study of stochastic processes is the martingale convergence theorem. Before stating this we need an understanding of the operation of conditioning with respect to a sub σ-field $\mathscr{G}$ on $(\Omega, \mathscr{F}, P)$, $\mathscr{G} \subset \mathscr{F}$. For a random variable X with finite expectation, the Radon–Nikodym theorem ensures the existence of a $\mathscr{G}$-measurable function $\phi(\omega)$ such that, for all $B \in \mathscr{G}$,

$$\int_B X \mathrm{d}P = \int_B \phi \mathrm{d}P$$

This function is called the conditional expectation of X given by $\mathscr{G}$, written $E\{X|\mathscr{G}\}$. Note that this is a random variable, rather than a constant. A sequence $\{X_n\}$ of random variables is called a martingale if, for $n > m$,

$$X_m = E\{X_n \mid \mathscr{G}_m\} \text{ a.s.}$$

where $\mathscr{G}_m$ is the smallest σ-field for which $X_1, X_2, \ldots, X_m$ are measurable. If $\{X_n\}$ is a martingale and Y a random variable such that

$$E\{Y|\mathscr{G}_n\} = X_n \text{ a.s.}$$

then there is a limit random variable X such that $X_n \to X$ a.s., and in mean. An early account of the theory of martingales is given in Reference 43, and there is a good summary of the important results in Chapter 7 of Tucker[38].

Martingales have devloped into a powerful tool in analysis. For example, many basic inequalities of Hardy and Littlewood concerning functions in L_p were known only for $p > 1$ or $p \geqslant 1$. The use of Brownian motion and martingales gives the correct formulation and proof for $0 < p < 1$. The fundamental work was done recently by Burkholder, Gundy and Silverstein, and there is an account by Garsia[45], which is the best place to start reading about it.

Markov processes form an important special class of stochastic processes. Intuitively a process is Markov if, given $X(t_0)$, the future development $X(t)$, $t > t_0$ and the past history $X(t)$, $t < t_0$ are independent. To study Markov processes for general state spaces and $T = [0, \infty]$ requires an elaborate measure-theoretic framework which can be set up in various ways. Most of the ideas can be thought of as generalisations of results known for Brownian motion – for a summary of this approach, see Chapter 12 of Breiman[35]. A functional analysis approach using semi-groups of operators and infinitesimal generators is given by Dynkin[46]. An important class of Markov processes in R^k were studied by Paul Lévy: they have increments which are independent. That is, for $t_1 < t_2 < \ldots < t_n$, the random vectors $\{X(t_i) - X(t_{i-1})\}$, $2 \leqslant i \leqslant n$ are independent with a distribution depending on the time increment $(t_i - t_{i-1})$. For a survey of the properties of such processes see Taylor in Reference 47. Markov processes with a countable state space can still have a complicated structure in continuous time. The best reference for this is Chung[48]. However, real simplifications result when not only is the state space discrete, but also time is replaced by the positive integers. Much of the probabilistic flavour of Markov processes can be obtained by studying discrete Markov chains – see Kemeny, Snell and Knapp[49] – or even random walks, where there is the additional condition of independent steps on a lattice in R^k: a full account of random walks was given by Spitzer[50].

Kakutani was the first to notice that a Brownian motion in R^3 has properties related to classical potential theory. If E is a closed body in R^3 and $x \notin E$, then the Brownian path starting at x either hists E or misses it.

$$\Phi(x, E) = P^x\{X(t) \in E \quad \text{for some} \quad t > 0\}$$

is harmonic in x with boundary values 1 on E and 0 at infinity so it is the solution of the exterior Dirichlet problem for E. Sets E which are hit with probability 0 from every starting point x correspond to sets of zero Newtonian capacity (for example, an infinite straight line). For *every* Markov process there is a corresponding potential theory. This was first proved by Hunt, but systematic accounts have been given by Meyer[51] and Blumenthal and Getoor[52]. The potential theory is simpler, but still interesting, for the special classes of Markov processes discussed in the last paragraph, and each of the books cited there discusses the relevant potential theory. Use of potential theory gives a 'fine topology' on the state space in which excessive functions (the appropriate generalisation of superharmonic functions) are continuous. It has recently been shown that classical results such as Picard's theorem for functions of a complex variable can be deduced from the properties of Brownian paths in the plane via potential theory.

Stochastic integrals and differentials have become an important tool in statistical mechanics. Because the square of the increment $(X(t + h) - X(t))$ on a Brownian path is of order h, if we consider the differential of a smooth function of the path $t \to X(t)$, we have to keep two terms in the power series expansion, and replace $(\mathrm{d}X)^2$ by $\mathrm{d}t$. Thus

$$\mathrm{d}f(X) = f'(X)\mathrm{d}X + \frac{1}{2} f''(X)\mathrm{d}X^2$$

or

$$\int_0^t f'(X)\mathrm{d}X = [f(X)] \int_0^t - \frac{1}{2} \int_0^t f''(X)\mathrm{d}s$$

A new theory is needed to take care of the extra term in the differential. This theory uses relatively simple properties of Brownian motion and martingales, but it requires its own development. There is a good elementary account in Chapter 5 of Lukacs[53] and a much fuller treatment in the book by McKean[54]. The 1975/76 Séminaire of the Institut de Mathématiques of Strasbourg University produced a new approach to stochastic integrals which is more general and at the same time easier to grasp.

Probability has been applied to many other branches of pure mathematics. As an example of diverse applications see the two books by Kac[55, 56].

REFERENCES

1. Lebesgue, H., 'Intégrale longueur, aire', *Annali di Matematica Pura ed Applicata*, 7, 231-358 (1902)
2. Lebesgue, H., *Leçons sur l'intégration et la recherche des fonctions primitives*, 2nd edn, Gauthier-Villars, Paris (1928)
3. Lebesgue, H., *Measure and the integral*, Holden-Day, San Francisco (1966)
4. Bourbaki, N., *Intégration*, Hermann, Paris (1965)
5. Saks, S., *Theory of the integral*, Hafner, New York (1950)
6. Fréchet, M., *La vie et l'oeuvre d'Emile Borel*, Monographies de l'Enseignement Mathématique 14, University of Geneva Press, Geneva (1965)
7. Sierpinski, W., *On the congruence of sets and their equivalence by finite decomposition*; reprinted Chelsea, New York (1967)
8. Munroe, M.E., *Measure and integration*, Addison-Wesley, Reading, Mass. (1953)
9. Taylor, S.J., *Introduction to measure and integration*, Cambridge University Press, Cambridge (1974)
10. Natanson, I.P., *Theory of functions of a real variable*, 2 vols, Ungar, New York (1955, 1960)
11. Benedetto, J.J., *Real variable and integration*, Teubner, Stuttgart (1976)
12. Royden, H.L., *Real analysis*, Macmillan, New York (1963)

13. Dunford, N. and Schwartz, J.T., *Linear operators, part I: general theory*, Interscience, New York (1958)
14. Ionescu-Tulcea, A. and Ionescu-Tulcea, C., *Topics in the theory of lifting*, Springer-Verlag, Berlin (1970)
15. Zygmund, A., *Trigonometric series*, Cambridge University Press, Cambridge (1959)
16. Katznelson, Y., *Introduction to harmonic analysis*, Wiley, New York (1968)
17. Hewitt, E. and Ross, K.A., *Abstract harmonic analysis*, Springer-Verlag, Berlin; Academic Press, New York (1963)
18. Halmos, P., *Lectures in ergodic theory*, Publications of the Mathematical Society of Japan 3, Nihon Sugakkai, Tokyo (1956)
19. Friedman, N.A., *Introduction to ergodic theory*, Van Nostrand, New York (1970)
20. Hewitt, E. and Stromberg, K.R., *Real and abstract analysis*, Springer-Verlag, Berlin (1965)
21. Riesz, F. and Sz.-Nagy, B., *Leçons d'analyse fonctionelle*, 5th edn, Akadémiai Kiadó, Budapest; Gauthier-Villars, Paris (1968)
22. Rogers, C.A., *Hausdorff measures*, Cambridge University Press, Cambridge (1970)
23. Rado, T., *Length and area*, Colloquium Publications 30, American Mathematical Society, New York (1948)
24. Lindahl, L.A. and Poulsen, F., *Thin sets in harmonic analysis*, Marcel Dekker, New York (1971)
25. Weir, A.J., *Lebesgue integration and measure*, Cambridge University Press, Cambridge (1973)
26. Dieudonné, J., *Treatise on analysis*, Academic Press, New York (1972)
27. Dinculeanu, N., *Vector measures*, Pergamon, Oxford (1967)
28. Fremlin, D.H., *Topological Riesz spaces and measure theory*, Cambridge University Press, Cambridge (1974)
29. Kolmogorov, A.N., *Foundations of the theory of probability*, Chelsea, New York (1956)
30. Gillies, D.A., *An objective theory of probability*, Methuen, London (1973)
31. Feller, W., *An introduction to probability theory and its applications*, Vol. 1, Wiley, New York (1966)
32. Lamperti, J., *Probability*, Benjamin, New York (1966)
33. Kingman, J.F.C. and Taylor, S.J., *Introduction to measure and probability*, Cambridge University Press, Cambridge (1974)
34. Bochner, S., *Harmonic analysis and the theory of probability*, University of California Press, Berkeley, Calif. (1955)
35. Breiman, L., *Probability*, Addison-Wesley, Reading, Mass. (1968)
36. Lévy, P., *Théorie de l'addition des variables aléatoires*, Gauthier-Villars, Paris (1954)
37. Gnedenko, B.V. and Kolmogorov, A.N., *Limit distributions for sums of independent random variables*, Addison-Wesley, Reading, Mass. (1954)
38. Tucker, H.G., *A graduate course in probability*, Academic Press, New York (1967)
39. Billingsley, P., *Convergence of probability measures*, Wiley, New York (1968)
40. Lévy, P., *Processus stochastiques et mouvement Brownien*, 2nd edn, Gauthier-Villars, Paris (1965)
41. Itô, K. and McKean, H.P., *Diffusion processes and their sample paths*, Academic Press, New York (1964)
42. Freedman, D., *Brownian motion and diffusion*, Holden-Day, San Francisco (1971)
43. Doob, J.L., *Stochastic processes*, Wiley, New York (1953)

44. Karlin, S., *A first course in stochastic processes,* Academic Press, New York (1966)
45. Garsia, A.M., *Martingale inequalities: seminar notes on recent progress,* Benjamin, Reading, Mass. (1973)
46. Dynkin, E.B., *Markov processes,* 2 vols, Springer-Verlag, Berlin (1965)
47. Kendall, D.G. and Harding, E.F. (eds.), *Stochastic analysis,* Wiley, London (1973)
48. Chung, K.L., *Markov chains with stationary transition probability,* Springer-Verlag, Berlin (1960)
49. Kemeny, J.G., Snell, J.L. and Knapp, A.W., *Denumerable Markov chains,* Van Nostrand, New York (1966)
50. Spitzer, F., *Principles of random walk,* Van Nostrand, New York (1964)
51. Meyer, P., *Probabilités et potentiel,* Hermann, Paris (1966)
52. Blumenthal, R.M. and Getoor, R.K., *Markov processes and potential theory,* Academic Press, New York (1968)
53. Lukacs, E., *Stochastic convergence,* Heath, Lexington (1968)
54. McKean, H.P., *Stochastic integrals,* Academic Press, New York (1969)
55. Kac, M., *Statistical independence in probability, analysis and number theory,* Carus Mathematical Monographs, Mathematical Association of America, Buffalo, N.Y. (1959)
56. Kac, M., *Probability and related topics in physical sciences,* Interscience, New York (1959)

11

Complex Analysis and Special Functions

I.N. Sneddon

11.1 BASIC TEXTS IN COMPLEX ANALYSIS

Since the theory of functions of a complex variable plays a central role in modern pure mathematics and in its applications to physics and engineering, there is naturally a plethora of elementary texts on the subject. Since there are many in English or in English-language editions, we shall restrict our attention to them. The choice of which to recommend is made difficult by the fact that some are designed for the needs of future pure mathematicians, some for students of engineering or the physical sciences and, as is to be expected, some try to 'capture both markets'.

For future mathematicians, either pure or applied, the best introductions are probably still the books by Copson[1] and Titchmarsh[2] although that by Ahlfors[3] is fast replacing them as the standard undergraduate textbook on complex analysis. The latter book carries with it all the authority of an author who has had a life-long career of distinguished research in function theory and this can be said too of the book by Nevanlinna and Paatero[4] and that by Levinson and Redheffer[5]. Although all these books reach the standard of rigour that would be expected of their authors, they would all — with the exception of that by Titchmarsh — be accessible to applied mathematicians and to graduate students of engineering. If one had to chose a text suitable for a class with 'mixed' interests, it would probably be the book by Levinson and Redheffer[5] which is beautifully written and has a wealth of illustrations of the use of the theory.

For the committed pure mathematician Saks and Zygmund[6] or the more recent book by Conway[7] would be more attractive. There are good books by Nehari[8] and by Duncan[9] which are suitable for a first course but none leads so directly or so easily into the deep theory of functions of a complex variable as does Conway's book[7]. A really fastidious student of pure mathematics will find that all of the books so far mentioned do not answer questions which are obvious to him — questions which concern the topology of the complex plane. Such a student should consult the books by Newman[10] and Whyburn[11] that are specifically concerned with such matters.

The choice for students of applied mathematics is just as bewildering. In many ways the book most suitable for people who wish to use complex function theory is one by MacRobert[12]. It has the added advantage that as illustrations of the general theory it develops those properties of the standard special functions which are of most frequent use in applications. The books by Churchill[13] and Dettman[14] are specifically written for 'users' of mathematics but in many ways they do not go far enough — despite their obvious virtues in providing a first course for such readers. There are several admirable Russian books available in English translation which are directed to this class of reader and which contain a wealth of examples of the use of complex function theory; the best of these are those by Smirnov[15] and by Fuchs and Levin and Fuchs and Shabat[16]. In this connection it should be observed that in complex function theory, as in other branches of mathematics, the beginner arrives at a true appreciation of the subject not only by reading the solution of problems but trying to solve problems for himself; in this connection he will find the book by Volkovyskii, Lunts and Aramanovich[17] particularly useful.

Some writers on the subject place the theory of functions of a complex variable in a wider context. Notable among them is Rudin, whose book[18] on real and complex analysis ranks as a modern classic. It is doubtful, however, if it would be a suitable text for someone with no previous knowledge of the subject. Mention should also be made of the (more elementary) books by Cartan[19] and Kaplan[20], which not only give a good account of the basic theory of functions of a single complex variable, but also have a chapter on functions of several complex variables, and of Chapter IX of Dieudonné's book[21], which, in a most interesting way, fits complex function theory into the framework of abstract analysis.

11.2 SPECIAL TOPICS IN COMPLEX ANALYSIS

There are many books which are devoted to the study of special classes of problems in complex analysis and of these most contain material for a first course in the subject also. Notable among those are the single volumes by Heins[22, 23], Fuchs[24] and Veech[25], the two-volume works by Carathéodory[26] and Hille[27] and the large three-volume ones by Markushevich[28] and Siegel[29].

One of the most fruitful applications of the theory is the use of Cauchy's integral theorem in the evaluation of integrals. This subject is discussed in most first texts on complex analysis – and is treated very fully in MacRobert[12] – but attention should be directed to the classic little monograph by Lindelof[30] and to the more recent one by Mitrinovic[31], both of which are devoted solely to this particular application.

The other concept of the theory which has proved to be of great use in applied mathematics is that of a conformal mapping, which is the name given to a mapping of the real plane into itself with the property that it maps intersecting smooth arcs into arcs which intersect at the same (oriented) angle. All elementary books give a proof that if a function f is holomorphic* in a domain D, then the mapping $f : D \to C$ is conformal but many interesting theoretical questions remain. These have been discussed by Bergman[32], Betz[33], Carathéodory[34], Courant[35], Nehari[8] and Jenkins[36]. In addition, the method is used extensively in classical field theory in physics to solve two-dimensional problems; the handbook by von Koppenfels and Stallman[37] and the dictionary of mappings by Kober[38] are invaluable in the solution of particular problems.

The idea of a Riemann surface is encountered early in the study of complex analysis but most elementary books are content to give only a simple example. The study of Riemann surfaces has, over the last 50 years, developed into a subject in its own right. The classic work is still that of Hermann Weyl[39] but more readable accounts – and describing more recent developments – have been given by Springer[40] and by Ahlfors and Sario[41]; the work by Pfluger[42] should also be consulted.

---

* It should be noted that the adjective now used to describe a differentiable function of a complex variable is *holomorphic;* in the recent past, the words *analytic* and *regular* were used but there seems to be a consensus of opinion in favour of *holomorphic.*

Functions which are holomorphic everywhere — the *entire* or *integral* functions — have been studied extensively; accounts of these investigations have been given by Valiron[43], Boas[44], Cartwright[45] and Holland[46]. A meromorphic function is one which is holomorphic everywhere except at poles; the best account of work on such functions has been given by Hayman[47].

In 1907 Koebe introduced the idea of a *schlicht* function. A function f is said to be schlicht in a domain D if for any two points z_1, z_2 of D, $f(z_1) = f(z_2)$ only if $z_1 = z_2$; nowadays, such functions are said to be *univalent*. Such functions have been studied extensively and the results reported by Montel[48], Schaeffer and Spencer[49] and Jenkins[36]. Multivalent functions have been discussed by Montel[48] and Hayman[50].

A problem which arises in control theory (and in other branches of applied mathematics where the Laplace transform is used) is that of deriving the location in the complex plane of the zeros of a polynomial in a single complex variable. The various methods of solution of this problem are described by Marsden[51]. The related problem of locating the critical points of holomorphic functions is treated by Walsh[52].

Recently there has been a revival of interest — particularly among numerical analysts — in the problem of approximating a holomorphic function by a polynomial or a rational function in the complex plane and the associated problem of interpolation in the complex plane. These problems have been discussed in the monographs by Whittaker[53], Walsh[54] and Sewell[55].

The classic work on Taylor series by Dienes[56] has not been superseded but the related monograph by Levinson[57] on gap and density theorems should be consulted.

In recent years there has been a revival of interest in the theory of automorphic functions but it is doubtful whether this should be classed as function theory or group theory. (It will be recalled that the classical theory of automorphic functions, created by Klein and Poincaré, was concerned with the study of analytic functions in the unit disc that are invariant under a discrete group of transformations.) The first comprehensive treatise in English is that of Ford[58] and it is still a valuable introduction to the subject; but the standard book on the subject is now that by Lehner[59] which, using modern algebraic and topological concepts, gives a concise but fairly detailed account of the important new ideas and techniques introduced in the three decades after the publication of Ford's book[58]. More recent developments are covered by Shimura[60].

One of the themes that has attracted much attention in functional analysis is that of demonstrating the value of interrelating the algebraic and analytic aspects of problems in function theory. The theory of linear spaces of holomorphic functions had its origins in the work on

H^p spaces* done in the 1920s by such mathematicians as Hardy, Littlewood, Privalov and Smirnow, who were primarily concerned with the properties of individual functions of the class H^p. In recent years the development of functional analysis has stimulated new interest in the H^p classes as linear spaces and this has provided new methods of attack, leading to important advances in the theory. An admirable book giving an account of both aspects of the subject — the classical and the modern — is that by Duren[61]. A slightly earlier monograph by Porcelli[62] on the same subject is also worth considering. In elementary complex analysis we study functions which map *D,* a domain in the complex plane C , into C itself. We generalise this concept by taking *D* to be a more general space — for example, a Banach algebra — and by defining holomorphic functions in this new context. Extensions of the classical theory obtained in this way are discussed in the monographs by Hoffman[63] and de Branges[64].

We may generalise complex function theory in another direction. We could regard the classical theory as being based on the concept of a holomorphic function as an ordered pair (u, v) of real-valued functions of two real variables x and y which satisfy the Cauchy–Riemann equations. The idea of defining a generalised holomorphic function as such an ordered pair when u and v are solutions of a first order system of elliptic differential equations was suggested independently by Bers and Vekua. A full account of the theory is given in Bers' lecture notes[65] and in Vekua's monograph[66].

A branch of complex analysis which falls between function theory and the theory of integral equations is the study of Cauchy integrals; this is concerned with the study of functions of the form

$$\int_L \frac{f(\tau)\mathrm{d}\tau}{z - \tau}$$

where L is an arc in the complex τ-plane and f need not be a holomorphic function of τ but obeys some such condition as a Lipschitz condition. An account of the properties of integrals of this kind has been given by Muskhelishvili[67].

Closely related to this topic and also of great value in applied mathematics is the Wiener-Hopf technique, which by the exploitation

---

* A function f which is holomorphic in the unit disc in the complex plane is said to be belong to the space $H^p(0 < p < \infty)$ if

$$M_p(r, f) = \left\{ \frac{1}{2\pi} \int_0^{2\pi} f \mid (r\ e^{i\theta}) \mid^p \mathrm{d}\theta \right\}^{1/p}$$

remains bounded as $r \to 1$; H^∞ is the class of bounded holomorphic functions in the unit disc.

of ideas from the theory of analytic continuation provides solutions of singular integral equations and of boundary value problems in the theory of partial differential equations. The standard work on this subject is the monograph by Noble[68].

11.3 FUNCTIONS OF SEVERAL COMPLEX VARIABLES

The first extended publication on the theory of functions of several complex variables would appear to be Osgood's Colloquium Lectures to the American Mathematical Society[69], the material of which was developed later as part of a textbook on function theory[70]. This was followed some ten years later by a book in German by Behnke and Thullen[71]. In all of these the motivation of theory sprang from a desire simple to generalise the classical theory of functions of a single complex variable, the development of which had occupied the attention of many distinguished mathematicians of the previous century. The book by Bochner and Martin[72] may well be described as following closely in the tradition of developing the subject for its own sake.

Recent activity has been motivated by developments in abstract analysis on the one hand and in more applied subjects such as quantum field theory and the theory of partial differential equations on the other.

Parts of the elementary theory comprising Hartog's theory, domains of holomorphy and automorphisms of bounded domains are treated in the books by Hörmander[73] and by Gunning and Rossi[74], and in the lecture notes of Cartan[75]. Introductory chapters occur in Cartan[19] and Kaplan[20], but by far the best introductions are provided by Kaplan[76] and Narasimhan[77].

Tbe local and global study of analytic sets and complex spaces is described in the books by Herve[78] and by Gunning and Rossi[74] and Narasimhan[79], and in the lecture notes by Cartan[80, 81] and Narasimhan[82].

The theory of Stein manifolds and of coherent analytic sheaves and global ideal theory are discussed in the books by Hörmander[73] and by Gunning and Rossi[74] and in the lecture notes by Cartan[75], Malgrange[83] and Frenkel[84].

In applications to quantum field theory interest is focused on the problem of analytic continuation of holomorphic functions of several complex variables and on the construction of the envelope of holomorphy of a function. The books by Fuks[85] and Vladimirov[86] are concerned with this and other applications of the theory; they also can be used as introductory texts, since they are lucidly written in a discursive style.

11.4 GENERAL TEXTS ON SPECIAL FUNCTIONS

At first sight the theory of the special functions of mathematical physics seems to be little more than a disorganised mass of formulae. There are nearly 50 special functions, each of which can be defined in a variety of ways; for each there is a profusion of integral and series representations, recurrence formulae and differential equations whose solutions are related functions, and so on. The very richness of the material and the challenge of imposing some kind of order in the apparent chaos has attracted some of the most outstanding mathematicians of the last 200 years to work – and write! – on the subject. Gauss, Euler, Fourier, Legendre, Bessel and Riemann are among the illustrious names appearing in the literature of the subject. In the words of Wigner: 'All of us have admired, at one time or another, the theory of the higher transcendental functions, also called special functions of mathematical physics. The variety of the properties of these functions, which can be expressed in terms of differential equations which they satisfy, in terms of addition theorems or definite integrals over the products of these functions, is truly surprising. It is surpassed only by the properties of the elementary transcendentals, that is the exponential function, and functions derived therefrom, such as the trigonometric functions. At the same time, special functions, as their full name already indicates, appear again and again as solutions of problems in theoretical physics.'

For this latter reason many introductory textbooks on special functions such as those by Hochstadt[87, 88], Lebedev[89], Rainville[90], Sneddon[91] and Spain and Smith[92] make use of rather elementary mathematical techniques. Since they are designed primarily for the use of applied mathematicians, they aim to introduce the reader to special functions by the use of the methods of elementary analysis and to provide them with sufficient background to make accessible to them numerous compilations of the properties of special functions, of which the most distinguished is that by Erdélyi *et al.*[93]

Since the special functions are holomorphic functions of their arguments it is usual to derive their more sophisticated properties by use of the theory of functions of a complex variable. The truth of this assertion is evidenced by the fact that the second half of Whittaker and Watson's classic treatise[94] on 'modern' analysis is devoted to the derivation of detailed properties of special functions by the systematic use of techniques developed in the first half. Similarly, in addition to compiling a collection of results which would be valuable to users of mathematics, Watson[95] stated in his monumental treatise on Bessel functions that his aim was to develop '... applications of the fundamental processes of the theory of functions of complex variables.

For this purpose Bessel functions are admirably adapted; while they offer at the same time a rather wider scope for the application of the parts of the theory of functions of a real variable than is provided by trigonometrical functions in the theory of Fourier series.' In a similar way, the best account of spherical and ellipsoidal harmonics is given by Hobson[96] who adopted Watson's philosophy.

When the unifying principle came it was, rather surprisingly, not from analysis, however, but from algebra − through the theory of group representations. In a historic paper in 1929 Eli Cartan first established the connection between special functions and group representations, and in the years immediately following its publication the application of the theory of group representations to quantum mechanics played a significant part in the exploitation of his discovery. For simple Lie groups, we can choose a basis in the representation space in such a way that the elements of some subgroup H are given by diagonal matrices whose elements are exponential functions. If h_1 and h_2 are members of H, the remaining members of the group can be represented in the form $h_1 g h_2$, where $g(t)$ is taken over a certain one-parameter manifold; it is then found that for particular groups the functions g can be identified with the special functions of mathematical physics. For instance, representations of the Euclidean group of the plane are connected with the Bessel functions of the first kind.

This method of developing the theory of special functions is certainly the best for serious students of pure mathematics, for it imparts meaning to what appears to be chaotic. Also, since simple group representation theory plays a vital role in modern applied mathematics, a student on the applied side would also derive profit as well as interest from approaching special functions from this direction. Fortunately there are now three good books in English, for this is the path of development taken by Vilenkin[97] in his encyclopaedic work and by Miller[98] in his more modest treatise. However, a more attractive introduction to this complex of ideas − especially for an applied mathematician − is provided by Talman[99] in his shorter book based on lectures by Wigner. It is not inappropriate to give Wigner the last word: 'Naturally, the common point of view from which the special functions are here considered, and also the natural classification of their properties, destroys some of the mystique which has surrounded and still surrounds, these functions. Whether this is a loss or a gain remains for the reader to decide.'

The great activity in the present century in functional analysis led to the study of special functions in the context of orthonormal bases in certain Hilbert spaces. This is the approach taken by Szego[100] in his classic work on orthogonal polynomials and more recently by Freud[101] and Sansone[102].

A different form of unified theory of special functions is due to

Truesdell[103]; this has been unfairly neglected, but the recent publication of McBride's monograph[104] shows that it has not been entirely forgotten. Truesdell's aim is 'to provide a general theory which motivates, discovers and coordinates such seemingly unconnected relations among particular special functions' as are known to exist. His method takes as its starting point the functional equation $\partial F(z, \sigma)/\delta z = F(z, \sigma + 1)$, to each of whose analytic solutions there corresponds the generating function of a set of special functions; he proceeds to show how the common formulae of the theory of special functions can be deduced from a handful of results derived from the study of this generating function.

Yet another approach is to regard Gauss' hypergeometric function and Kummer's confluent hypergeometric function as forming the heart of the theory of special functions, since many special functions can be expressed in terms of one or the other. Natural generalisations of these functions are the generalised hypergeometric function $_pF_q$, MacRobert's E-function and Meijer's G-function, and it is possible to set up a corpus of results relating to these functions and then to deduce the properties of the classical special functions from it. This approach, which seems to the present writer to be singularly lacking in interest, is the one taken by Luke[105]; his book cannot be recommended as a textbook but should prove useful as a reference book.

The asymptotic expansions of special functions play a significant role both in the theory and in its applications. For that reason, reference must be made to the excellent book by Olver[106], which develops the general theory of asymptotic behaviour hand-in-hand with the theory of special functions.

11.5 SPECIAL TOPICS IN THE THEORY OF SPECIAL FUNCTIONS

The theory of Legendre functions and the associated theory of spherical harmonics is described in detail by Hobson[96] and by MacRobert[107], although, as has been pointed out already, many important formulae for Legendre and associated Legendre functions are to be found in MacRobert[12].

Watson's treatise[95] remains, after more than half-a-century, the best book on Bessel functions, but a student of applied mathematics looking for an elementary introduction might well be advised to begin with Tranter's book[108].

The Chebyshev* polynomials are usually treated in general texts on

---

* This name has been variously — and often quite bizarrely — transliterated from Cyrillic characters but the commonest forms are Chebyshev and Tchebycheff.

special functions, but because of their importance in approximation theory and numerical analysis in recent years specialised texts have been devoted to them, of which the foremost are those by Karlin and Studden[109], Fox and Parker[110] and Rivlin[111].

Since an elliptic function is a doubly periodic meromorphic function, elementary results concerning elliptic functions are more usually found in books on complex analysis than in those on special functions. The best overall introduction is probably provided by the relevant chapters of MacRobert[12] or of Whittaker and Watson[94]. Jacobian elliptic functions and elliptic integrals are discussed in detail in Tricomi[112] and Neville[113]. Tricomi's book also contains a discussion of the properties of Weierstrass' elliptic functions and related functions. Siegel[29] discusses elliptic functions and uniformisation theory in vol. 1 of his treatise, while in vols 2 and 3 he discusses automorphic functions and Abelian integrals, and Abelian functions and modular functions of several variables, respectively. Graeser's monograph[114] in German is also a valuable source of material. A useful guide to the theory of both elliptic functions and elliptic integrals is provided by the handbook by Byrd and Friedman[115]; this also contains a valuable collection of formulae.

Mathieu, Lamé and allied functions are treated from the standpoint of solutions of differential equations with periodic coefficients by Arscott[116]. A brief introduction to the properties of Mathieu functions is contained in Whittaker and Watson's treatise[94], and fuller accounts are given in the books by McLachlan[117], Campbell[118] and Meixner and Schafke[119].

Confluent hypergeometric functions are treated in detail by Buchholz[120] and generalised hypergeometric functions by Slater[121].

Other special functions, such as those associated with the names of Hermite and Laguerre, which arise in the solution of special problems in wave mechanics are treated in most of the general texts to which reference has been made above but also in larger treatises on the methods of mathematical physics such as that by Morse and Feshbach[122].

REFERENCES

1. Copson, E.T., *An introduction to the theory of functions of a complex variable,* Clarendon Press, Oxford (1935)
2. Titchmarsh, E.C., *The theory of functions,* 2nd edn, Oxford University Press, Oxford (1939)
3. Ahlfors, L.V., *Complex analysis,* rev. edn, McGraw-Hill, New York (1966)
4. Nevanlinna, R. and Paatero, V., *Introduction to complex analysis,* Addison-Wesley, Reading, Mass. (1969)

5. Levinson, N. and Redheffer, R.M., *Complex variables*, Holden-Day, San Francisco (1970)
6. Saks, S. and Zygmund, A., *Analytic functions*, 3rd edn, PWN, Warsaw (1971)
7. Conway, J.B., *Functions of one complex variable*, Springer-Verlag, Berlin (1973)
8. Nehari, Z., *Conformal mapping*, McGraw-Hill, New York (1959)
9. Duncan, J., *Elements of complex analysis*, Wiley, New York (1968)
10. Newman, M.H.A., *Elements of the topology of plane sets of points*, 2nd edn, Cambridge University Press, Cambridge (1951)
11. Whyburn, G.T., *Topological analysis*, rev. edn, Princeton University Press, Princeton (1964)
12. MacRobert, T.M., *Functions of a complex variable*, 5th edn, Macmillan, London (1962)
13. Churchill, R.V., *Complex variables and applications*, McGraw-Hill, New York (1960)
14. Dettman, J.W., *Applied complex variables*, Macmillan, New York (1965)
15. Smirnov, V.I., *Complex variables; special functions*, Pergamon, Oxford (1964)
16. Fuchs, B.A. *et al.*, *Functions of a complex variable, and some of their applications*, 2 vols, Pergamon Press, Oxford (1961, 1964)
17. Volkovyskii, L.I., Lunts, G.L. and Aramanovich, V.S., *A collection of problems on complex analysis*, Pergamon Press, Oxford (1965)
18. Rudin, W., *Real and complex analysis*, McGraw-Hill, New York (1966)
19. Cartan, H., *Elementary theory of analytic functions of one or several complex variables*, Addison-Wesley, Reading, Mass. (1963)
20. Kaplan, W., *Introduction to analytic functions*, Addison-Wesley, Reading, Mass. (1966)
21. Dieudonné, J., *Foundations of modern analysis*, enl. edn, Academic Press, New York (1969)
22. Heins, M., *Selected topics in the classical theory of functions of a complex variable*, Holt, Rinehart and Winston, New York (1962)
23. Heins, M. *Complex function theory*, Academic Press, New York (1968)
24. Fuchs, W.H.J., *Topics in the theory of functions of one complex variable*, Van Nostrand, Princeton (1967)
25. Veech, W.A., *A second course in complex analysis*, Benjamin, New York (1967)
26. Carathéodory, C., *Theory of functions*, 2 vols, Chelsea, New York (1964)
27. Hille, E., *Analytic function theory*, 2 vols, Ginn, Boston (1959)
28. Markushevich, A.I. *Theory of functions of a complex variable*, 3 vols, Prentice-Hall, Englewood Cliffs, N.J. (1967)
29. Siegel, C.L., *Topics in complex function theory*, 3 vols, Wiley, New York (1969-1973)
30. Lindelof, E., *Le calcul des residus*, Gauthier-Villars, Paris (1905)
31. Mitrinovič, D.S., *Calculus of residues*, Noordhoff, Groningen (1966)
32. Bergman, S., *The kernel function and conformal mapping*, Mathematical Surveys 5, American Mathematical Society New York (1950)
33. Betz, A., *Konforme Abbildung*, 2nd edn, Springer-Verlag, Berlin (1964)
34. Carathéodory, C., *Conformal representations*, 2nd edn, Cambridge University Press, Cambridge (1952)
35. Courant, R., *Dirichlet's principle, conformal mapping and minimal surfaces*, Interscience, New York (1950)
36. Jenkins, J.A., *Univalent functions and conformal mapping*, 2nd edn, Springer-Verlag, Berlin (1964)
37. von Koppenfels, W. and Stallman, F., *Praxis der konformen Abbildung*, Springer-Verlag, Berlin (1959)

38. Kober, H., *Dictionary of conformal representations,* 2nd edn, Dover, New York (1957)
39. Weyl, H., *The concept of a Riemann surface,* Addison-Wesley, Reading, Mass. (1964)
40. Springer, G., *Introduction to Riemann surfaces,* Addison-Wesley, Reading, Mass. (1957)
41. Ahlfors, L.V. and Sario, L., *Riemann surfaces,* Princeton University Press, Princeton (1960)
42. Pfluger, A., *Theorie der Riemannschen Flächen,* Springer-Verlag, Berlin (1957)
43. Valiron, G., *Lectures on the general theory of integral functions,* the author, Toulouse (1923)
44. Boas, R.P., *Entire functions,* Academic Press, New York (1954)
45. Cartwright, M.L., *Integral functions,* Cambridge University Press, Cambridge (1962)
46. Holland, A.S.B., *Entire functions,* Academic Press, New York (1973)
47. Hayman, W.K., *Meromorphic functions,* Clarendon Press, Oxford (1964)
48. Montel, P., *Leçons sur les fonctions univalentes ou multivantes,* Gauthier-Villars, Paris (1933)
49. Schaeffer, A.C. and Spencer, D.C., *Coefficient regions for schlicht functions,* Colloquium Publications 35, American Mathematical Society, New York (1950)
50. Hayman, W.K., *Multivalent functions,* Cambridge University Press, Cambridge (1958)
51. Marden, M., *The geometry of zeros of a polynomial in a complex variable,* Mathematical Surveys 3, American Mathematical Society, New York (1949)
52. Walsh, J.L., *The location of critical points of analytic and harmonic functions,* Colloquium Publications 34, American Mathematical Society, New York (1950)
53. Whittaker, J.M., *Interpolatory function theory,* Cambridge University Press, Cambridge (1960)
54. Walsh, J.L., *Interpolation and approximation by rational functions in the complex domain,* Colloquium Publications 20, American Mathematical Society, New York (1960)
55. Sewell, W.E., *Degree of approximation of polynomials in the complex domain,* Princeton University Press, Princeton (1942)
56. Dienes, P., *The Taylor series: an introduction to the theory of functions of a complex variable,* Clarendon Press, Oxford (1931)
57. Levinson, N., *Gap and density theorems,* Colloquium Publications 26, American Mathematical Society, New York (1940)
58. Ford, L.R., *Automorphic functions,* McGraw-Hill, New York (1929)
59. Lehner, J., *Discontinuous groups and automorphic functions,* Mathematical Surveys 8, American Mathematical Society, Providence, R.I. (1964)
60. Shimura, G., *Introduction to the arithmetic theory of automorphic functions,* Princeton University Press, Princeton (1971)
61. Duren, P.L., *Theory of H^p spaces,* Academic Press, New York (1970)
62. Porcelli, P., *Linear spaces of analytic functions,* Rand McNally, Chicago (1966)
63. Hoffman, K., *Banach spaces of analytic functions,* Prentice-Hall, Englewood Cliffs, N.J. (1962)
64. de Branges, L., *Hilbert spaces of entire functions,* Prentice-Hall, Englewood Cliffs, N.J. (1968)
65. Bers, L., *Theory of pseudo-analytic functions,* Lecture Notes, New York University, New York (1953)

66. Vekua, I.N., *Generalized analytic functions*, Pergamon Press, Oxford (1962)
67. Muskhelishvili, N.I., *Singular integral equations*, Noordhoff, Groningen (1953)
68. Noble, B., *Methods based on the Wiener–Hopf technique for the solution of partial differential equations*, Pergamon Press, Oxford (1958)
69. Osgood, W.F., *Topics in the theory of functions of several complex variables*, Colloquium Publications 4 Part II, American Mathematical Society, New York (1914)
70. Osgood, W.F., *Lehrbuch der Funktionentheorie*, Bd II, Teubner, Leipzig (1924)
71. Behnke, H. and Thullen, P., *Theorie der Funktionen mehrer komplexer Veränderlichen*, Springer-Verlag, Berlin (1934); reprinted Chelsea, New York
72. Bochner, S. and Martin, W.T., *Several complex variables*, Princeton University Press, Princeton, N.J. (1948)
73. Hörmander, L., *An introduction to complex analysis in several variables*, Van Nostrand, Princeton (1966)
74. Gunning, R.C. and Rossi, H., *Analytic functions of several complex variables*, Prentice-Hall, Englewood Cliffs, N.J. (1965)
75. Cartan, H., *Séminaire 1951/52*, École Normale Supérieure, Paris (1952)
76. Kaplan, W., *Functions of several complex variables*, Ann Arbor Publishers, Ann Arbor, Mich. (1964)
77. Narasimhan, R., *Several complex variables*, University of Chicago Press, Chicago (1971)
78. Herve, M., *Several complex variables: local theory*, Oxford University Press, Oxford (1963)
79. Narasimhan, R., *Analysis on real and complex manifolds*, North-Holland, Amsterdam (1968)
80. Cartan, H., *Séminaire 1953/54*, École Normale Supérieure, Paris (1954)
81. Cartan, H., *Séminaire 1960/61*, École Normale Supérieure, Paris (1961)
82. Narasimhan, R., *Introduction to the theory of analytic spaces*, Lecture Notes in Mathematics 25, Springer-Verlag, Berlin (1966)
83. Malgrange, B., *Lectures on functions of several complex variables*, Tata Institute, Bombay (1958)
84. Frenkel, J., *Séminaire sur les théorèmes A et B pour les éspaces de Stein*, Université de Strasbourg (Institut de Mathématiques), Strasbourg (1965)
85. Fuks, B.A., *Introduction to the theory of analytic functions of several complex variables*. Translations of Mathematical Monographs 8, American Mathematical Society, Providence, R.I. (1963)
86. Vladimirov, V.S., *Methods of the theory of functions of many complex variables*, M.I.T. Press, Cambridge, Mass. (1966)
87. Hochstadt, H., *Special functions of mathematical physics*, Holt, Rinehart and Winston, New York (1961)
88. Hochstadt, H., *The functions of mathematical physics*, Wiley, New York (1971)
89. Lebedev, N.N., *Special functions and their applications*, Prentice-Hall, Englewood Cliffs, N.J. (1965)
90. Rainville, E.D., *Special functions*, Macmillan, New York (1960)
91. Sneddon, I.N., *The special functions of mathematical physics and chemistry*, 2nd edn, Oliver and Boyd, Edinburgh (1961)
92. Spain, B. and Smith, M.G., *Functions of mathematical physics*, Van Nostrand Reinhold, London (1970)
93. Erdélyi, A. et al., *Higher transcendental functions*, 3 vols, Macmillan, New York (1953-1955)

94. Whittaker, E.T. and Watson, G.N., *A course of modern analysis*, Cambridge University Press, Cambridge (1927)
95. Watson, G.N., *A treatise on the theory of Bessel functions*, 2nd edn, Cambridge University Press, Cambridge (1944)
96. Hobson, E.W., *The theory of spherical and ellipsoidal harmonics*, Cambridge University Press, Cambridge (1931)
97. Vilenkin, N. Ja., *Special functions and the theory of group representations*, Translations of Mathematical Monographs 22, American Mathematical Society, Providence, R.I. (1968)
98. Miller, W., *Lie theory and special functions*, Academic Press, New York (1968)
99. Talman, J.D., *Special functions: a group theoretic approach*, Benjamin, New York (1968)
100. Szego, G., *Orthogonal polynomials*, 3rd edn, Colloquium Publications 23, American Mathematical Society, Providence, R.I. (1967)
101. Freud, G., *Orthogonal polynomials*, Pergamon Press, Oxford (1971)
102. Sansone, G., *Orthogonal functions*, rev. edn, Interscience, New York (1959)
103. Truesdell, C.A., *An essay towards a unified theory of special functions*, Princeton University Press, Princeton, N.J. (1948)
104. McBride, E.B., *Obtaining generating functions*, Springer-Verlag, Berlin (1971)
105. Luke, Y.L., *The special functions and their approximations*, 2 vols, Academic Press, New York (1969)
106. Olver, F.W.J., *Asymptotics and special functions*, Academic Press, New York (1974)
107. MacRobert, T.M., *Spherical harmonics*, 3rd edn, Pergamon, Oxford (1967)
108. Tranter, C.J., *Bessel functions with some physical applications*, English Universitites Press, London (1968)
109. Karlin, S. and Studden, W.J., *Tchebycheff systems: with applications in analysis and statistics*, Wiley, New York (1966)
110. Fox, L. and Parker, I.B., *Chebyshev polynomials in numerical analysis*, Oxford University Press, Oxford (1968)
111. Rivlin, T.T., *The Chebyshev polynomials*, Wiley, New York (1974)
112. Tricomi, F.G., *Elliptische Funktionen*, Akademische Verlag, Leipzig (1948)
113. Neville, E.H., *Jacobian elliptic functions*, 2nd edn, Clarendon Press, Oxford (1951)
114. Graeser, E., *Einführung in die Theorie der Elliptischen Funktionen und deren Anwendung*, Oldenbourg, Munich (1950)
115. Byrd, P.F. and Friedman, M.D., *Handbook of elliptic integrals for engineers and physicists*, Springer-Verlag, Berlin (1954)
116. Arscott, F.M., *Periodic differential equations*, Pergamon Press, Oxford (1964)
117. McLachlan, N.W., *Theory and applications of Mathieu functions*, Clarendon Press, Oxford (1947)
118. Campbell, R., *Théorie générale de l'équation de Mathieu*, Masson et Cie, Paris (1955)
119. Meixner, J. and Schafke, F.W., *Mathieusche Funktionen und Sphäroidfunktionen*, Springer-Verlag, Berlin (1954)
120. Buchholz, H., *The confluent hypergeometric function with emphasis on its application*, Springer-Verlag, Berlin (1969)
121. Slater, L.J., *Generalized hypergeometric functions*, Cambridge University Press, Cambridge (1966)
122. Morse, P.M. and Feshbach, H., *Methods of theoretical physics*, McGraw-Hill, New York (1953)

12

Convexity

P. McMullen

12.1 INTRODUCTION

One of the main problems in trying to survey the literature of convexity is the very diversity of the subject; indeed, while certain areas have been investigated for many years, and have acquired a considerable theory as a result, others have been studied for as long, and yet still remain a collection of only vaguely connected results. Another very severe problem is that since the book of Bonnesen and Fenchel[1], there has been no attempt to cover the whole subject even of finite dimensional convexity in a single text. In fact, even Reference 1 contains little material of a combinatorial nature.

At this point, a word of explanation is in order. Although much work has been done in recent years on convexity in infinite dimensional spaces, which generalises finite dimensional results, and is even on occasion quite combinatorial, nevertheless it largely remains true that infinite dimensional convexity is a tool for use in functional analysis and measure theory. For this reason, and even more importantly because of personal interest, we shall confine our attention here to the finite dimensional theory. Indeed, of necessity there has had to be a selection of material, and this selection has been guided by the same personal interest, though we have tried to give a broad view of the subject. However, we have possibly overemphasised the combinatorial material at the expense of the rest.

We begin by listing a number of especially useful texts. From the historical point of view, the pioneer of convexity is Minkowski[2] (the reference is to his complete works); we should not, however, neglect the earlier work of Steiner[3], Schläfli[4], Eberhard[5] or Brückner[6]. We have mentioned Bonnesen-Fenchel[1]; among other important earlier works are Blaschke[7,8], Steinitz[9] and Steinitz and Rademacher[10]. As far as more recent texts are concerned, for the general and metrical theory, we make most references to Eggleston[11], Hadwiger[12], Rockafellar[13] and Valentine[14]. Other particularly useful books are Aleksandrov[15], Busemann[16], Hadwiger[17] and Santaló[18]. The greatest relative increase in material in latter years has been in the theory of convex polytopes. The indispensible monograph here is Grünbaum[19]; we also frequently refer to the introductory text McMullen and Shephard[20]. Other works on polytopes are Aleksandrov[21], Fejes Tóth[22] and Lyusternik[23], although here again metrical interests predominate. For more recent developments, the survey articles Grünbaum and Shephard[24] and Grünbaum[25] are very useful.

For a much broader view of the subject than space here permits, the reader may also find it worthwhile to consult the two collections of articles edited by Klee[26] and Fenchel[27], which are based on conferences.

For organisational convenience, we have divided the discussion into Basic theory, Combinatorial theory and Metrical theory, although these divisions can in no way be regarded as watertight.

12.2 BASIC THEORY

12.2.1 Convex sets

A subset K of a real linear space is *convex* if, for every $x, y \in K$, the (closed) *line segment*

$$[xy] = \{(1-\lambda)x + \lambda y \,|\, 0 \leqslant \lambda \leqslant 1\}$$

is contained in K. In this section, we shall survey the basic properties of convex sets; as we said in the introduction, we shall confine our attention to finite dimensional spaces. In fact, we shall work in d-dimensional Euclidean space E^d, which is endowed with the usual inner product $\langle . , . \rangle$ and norm or distance $\| . \|$.

The algebraic theory of convex sets has many points in common with the theory of linear spaces (see References 1, 11, 13, 14, 19, 20).

For example, if K, K_1 and K_2 are convex subsets of E^d, in which we shall work, and if λ is real, then the *scalar multiple*

$$\lambda K = \{\lambda x | x \in K\}$$

and the *vector* or *Minkowski sum*

$$K_1 + K_2 = \{x_1 + x_2 | x_2 \in K_1, x_2 \in K_2\}$$

are also convex. The intersection of convex sets is convex. The convex hull conv$X$ of a subset X of E^d is the intersection of all convex sets containing X; it is also the set of all *convex combinations*

$$\sum_{i=1}^{k} \lambda_i x_i \quad \left(\lambda_i \geqslant 0, \sum_{i=1}^{k} \lambda_i = 1\right)$$

of points $x_i \in X$ ($i = 1, \ldots, k$). Every convex set K has a natural *dimension* dim$K$, which is the dimension of the smallest affine subspace (linear variety, flat) aff$K$, the *affine hull* of K, which contains K.

However, there is a difference when we come to Carathéodory's theorem. If $y \in$ conv$X$, then y is a convex combination of at most $d + 1$ points of X; the points of X picked out, however, will in general vary with y, so we do not have an exactly analogous theory of 'convex bases'. There are various refinements of Carathéodory's theorem (Reay[28]; see also References 11, 14); a generalisation occurs in Valentine[14].

A closely connected result is Radon's theorem: a set X of at least $d + 2$ points in E^d admits a partition into disjoint subsets Y and Z, such that conv$Y \cap$ conv$Z \neq \phi$. There are a number of generalisations of Radon's theorem, with conditions ensuring that dim(conv$Y \cap$ conv$Z) \geqslant k$, or enabling one to partition X into more subsets (Tverberg[29], Reay[30]).

A third result, closely associated with the previous two, is Helly's theorem. This states that if every $d + 1$ members of a finite family of convex sets in E^d has a non-empty intersection, then so has the whole family. Again there are many generalisations, and many Helly-type theorems are known; we refer to References 11, 14 and to Danzer, Grünbaum and Klee[31] for an extensive discussion. For a stronger version of Helly's theorem, involving infinitely many convex sets, see Reference 13. For some of the relationships between the theorems of Helly and Carathéodory, see Reference 11.

As a final algebraic comment, we remark that convexity is preserved by affine maps, and even by suitable projective maps.

The topological properties of convex sets are fairly simple. We work with the Euclidean metric topology, but recall that finite dimensional real linear spaces are topological linear spaces in a unique way. The basic facts are that the closure cl$K$ and interior int$K$ of a convex set K

are convex. It is often more convenient to deal with a convex set K within aff$K$; thus the *relative interior* relint$K$ is the interior of K relative to aff$K$. So relint$K$ is convex; in fact, if $x \in$ cl$K$ and $y \in$ relint$K$, then every point of the open line segment

$$]xy[= \{(1-\lambda)x + \lambda y \,|\, 0 < \lambda < 1\}$$

lies in relint$K$. One further has cl$K$ = cl(relint$K$) and relint$K$ = relint(cl$K$). We denote boundary and relative boundary by bd and relbd. For details see References 13 and 14.

Carathéodory's theorem leads to the fact that the convex hull of a compact set is compact.

Before proceeding, let us note particular examples of convex sets. A *polytope* is the convex hull of a finite set; the family of polytopes in E^d is denoted $\mathscr{P}^d$. A *polyhedral set* is the intersection of a finite number of closed half-spaces — that is, sets of the form

$$H^-(u, \alpha) = \{x \in E^d \,|\, \langle x, u \rangle \leqslant \alpha\};$$

the polytopes are precisely the bounded polyhedral sets. C is a (convex) *cone* with *apex a* if, for all $x \in C$, the half-line or *ray*

$$[ax = \{(1-\lambda)a + \lambda x \,|\, \lambda \geqslant 0\}$$

lies in C. The family of compact convex sets in E^d is denoted $\mathscr{K}^d$; if $K \in \mathscr{K}^d$ has non-empty interior, we call K a *convex body*. The *unit ball* $B^d = \{x \in E^d \,|\, \|x\| \leqslant 1\}$ is a convex body; its boundary is the *unit sphere* $S^{d-1} = \{x \in E^d \,|\, \|x\| = 1\}$, which is not convex.

With every convex set K is associated a cone, its *recession cone* (asymptotic cone, characteristic cone) rec$K$, defined to be

$$\mathrm{rec}K = \cap\{\lambda(K-z) \,|\, \lambda > 0\}$$

where $z \in$ relint$K$ is arbitrarily chosen. The face of apices of rec$K$ is the *lineality space* of K, rec$K \cap (-\mathrm{rec}K)$. Every closed convex set can be expressed as a direct sum of its lineality space and a line-free convex set.

A hyperplane

$$H(u, \alpha) = \{x \in E^d \,|\, \langle x, u \rangle = \alpha\}$$

is said to *support* a convex set K, with *outer normal u*, if

$$\alpha = \sup \{\langle x, u \rangle \,|\, x \in K\} = h(K, u)$$

(say). Every boundary point of K belongs to at least one support hyperplane of K; there are many proofs of this result (see References 11, 13, 14, 19, 20). A closed convex set K is *smooth* if each boundary point of K lies in exactly one support hyperplane. K is *strictly convex* if each support hyperplane meets K in a single point; equivalently, if x, $y \in K$, then $]xy[\subseteq \text{int}K$. If both hold, then K is *regular* (although this word is used differently by different authors), and there is a one to one map between bd$K$ and S^{d-1}, which is continuous both ways. More generally, the spherical image of a subset A of bd$K$ is the set of unit outer normals to support hyperplanes of K which meet A, and the inverse spherical image of a subset ω of S^{d-1} is the set of points of bd$K$ lying in support hyperplanes of K with outer normals in ω (Busemann[16]). As a consequence of the existence of support hyperplanes, every closed convex set is the intersection of closed half-spaces.

The function $h(K,.)$ introduced above is the *support function* of K; it is convex (cf. subsection 12.2.2) and positive homogeneous, so that

$$h(K,\lambda u + \mu v) \leqslant \lambda h(K,u) + \mu h(K, v)$$

for $u, v \in E^d$ and $\lambda, \mu \geqslant 0$. Any convex positive homogeneous function is the support function of a closed convex set; three proofs are known, the first uses differentiability properties of convex functions (cf. subsection 12.2.2), the second uses polarity (see below), while the third and most recent is elementary (see References 11, 13, 14, McMullen[32]).

Two convex sets K_1 and K_2 are *separated* by the hyperplane $H(u, \alpha)$ if they lie in the different half-spaces bounded by $H(u, \alpha)$, so that (say) $\langle x_1, u \rangle \leqslant \alpha \leqslant \langle x_2, u \rangle$ for all $x_1 \in K_1$ and $x_2 \in K_2$. If we have strict inequality, the separation is *strict*. Support and separation properties can be deduced one from the other; for details see References 11, 13, 14, 19.

If X is any subset of E^d, then its *polar* X^* is defined by

$$X^* = \{x^* \in E^d | \langle x, x^* \rangle \leqslant 1 \quad \text{for all} \quad x \in X\}$$

X^* is closed and convex, and $X^{**} = \text{cl conv}(X \cup \{o\})$. If C is a cone with apex o, then so is C^*, and the definition in this case can be replaced by

$$C^* = \{x^* \in E^d | \langle x, x^* \rangle \leqslant 0 \quad \text{for all} \quad x \in C\};$$

if C is closed and convex, then $C^{**} = C$. If $K \in \mathscr{K}^d$ with $o \in \text{int}K$, then $K^* \in \mathscr{K}^d$ with $o \in \text{int}K^*$, and $K^{**} = K$. In this case, moreover, the support function of K^* is the *distance* or *gauge function* $g(K,.)$ of K, defined by

$$g(K, x) = \inf\{\lambda \geqslant 0 | x \in \lambda K\}$$

If K is *centrally symmetric* (about o) – that is, $K = -K$ – then $g(K,.)$ is a norm. The metric geometry obtained by using such a norm is a *Minkowski geometry*; for an introduction to these, see Minkowski[2] and Busemann[33]. For general details, see References 11, 13, 14, 19, 20.

The family of compact subsets of E^d has a metric, the *Hausdorff metric* ρ, defined by

$$\rho(X, Y) = \inf\{\rho \geqslant 0 \mid X \subseteq Y + \rho B^d, Y \subseteq X + \rho B^d\}$$

In this metric the closed subsets of any fixed compact set themselves form a sequentially compact set; this is the Blaschke selection theorem. The limit of a convex set is again convex; thus $\mathscr{K}^d$ is a locally compact metric space in the Hausdorff metric. One observes that

$$\rho(K_1, K_2) = \sup\{|h(K_1, u) - h(K_2, u)| \mid u \in S^{d-1}\}$$

for $K_1, K_2 \in \mathscr{K}^d$. For details see References 1, 11, 12, 14.

Many functions on $\mathscr{K}^d$ one naturally considers are continuous in the Hausdorff metric. Thus it is often useful first to prove results for a suitable dense subset of $\mathscr{K}^d$. One such dense subset is the family $\mathscr{P}^d$ of polytopes. Another consists of the regular convex sets. In fact, one can take the support functions of the approximating bodies to be analytic (Minkowski[2]), or even algebraic (Hammer[34]; a particularly short proof is due to Firey[35]).

We conclude this section by remarking that sets can be characterised as convex by means of local convexity or local support properties; see Reference 14 for a discussion of these.

12.2.2   Convex functions

A real-valued function f whose domain is a convex subset D of E^d is called *convex* if

$$f((1-\lambda)x + \lambda y) \leqslant (1-\lambda) f(x) + \lambda f(y)$$

for all $x, y, \in D$ and $0 \leqslant \lambda \leqslant 1$. If $-f$ is convex, we call f *concave*. The connection between convex sets and convex functions is intimate, for f is a convex function if and only if its *epigraph*

$$\mathrm{epi} f = \{(x, \lambda) \mid \lambda \geqslant f(x), x \in D\}$$

is convex. It is often convenient to consider extended real functions, which can also take the values $\pm\infty$; the *effective domain* of f is then

$$\mathrm{dom} f = \{x \in E^d \mid f(x) < +\infty\}$$

which is a convex set. So, for example, to a convex set K corresponds its *indicator function* $\delta(K, .)$, defined by

$$\delta(K, x) = \begin{cases} 0, & \text{if } x \in K \\ +\infty, & \text{if } x \notin K \end{cases}$$

Convex functions have nice continuity properties. A convex function f is upper semicontinuous – that is, if

$$\lim_{i \to \infty} x_i = x,$$

then

$$\lim_{i \to \infty} f(x_i) \leqslant f(x).$$

So, continuity is equivalent to lower semicontinuity, and this, in turn, is equivalent to the epigraph of f being closed. This leads to a *closure operation* for convex functions: clf is defined by epi(clf) = cl(epif). In any case, f and clf coincide on relint(domf). (For details see Reference 13.)

Convex functions possess *directional derivatives*

$$f'(x;y) = \lim_{\lambda \downarrow 0} (f(x + \lambda y) - f(x))/\lambda$$

whenever $f(x)$ is finite (the limit may be $+\infty$). Then $f'(x;y)$ is a positive homogeneous convex function of y. Thus $-f'(x; -y) \leqslant f'(x;y)$; if we have equality, then f has a derivative in direction y. If f has derivatives at x in d independent directions, then f is differentiable at x, and $f'(x;y) = \langle \nabla f(x), y \rangle$, where $\nabla f(x)$ is the gradient of f at x. But we have a concept which replaces the gradient if f is not differentiable. A vector x^* is a *subgradient* of f at x if $f(z) \geqslant f(x) + \langle x^*, z - x \rangle$ for all z. This means that $(x^*, -1)$ is an outer normal to a support hyperplane of epif at $(x, f(x))$. The set of all subgradients of f at x forms the *subdifferential* $\partial f(x)$, which is a convex set. (For the above, see References 1 and 13.)

In fact, a convex function on an open set is differentiable, and, indeed, twice differentiable, almost everywhere (Busemann and Feller[36], Aleksandrov[37]).

An important idea, introduced by Fenchel, is that of conjugate functions. If f is a convex function, its *conjugate* f^* is defined by

$$f^*(x^*) = \sup \{\langle x, x^* \rangle - f(x) | x \in E^d\}$$

Then f^* is closed and convex, being the supremum of affine functions, and $f^{**} = $ clf. As an example, the conjugate of the indicator function of a closed convex set is its support function. A connection with subdifferentials is that, if $f(x) = $ clf(x), then $x^* \in \partial f(x)$ if and only if $x \in \partial f^*(x^*)$. This idea is closely related to the Legendre transform. For details of the above, see Reference 13.

12.3   COMBINATORIAL THEORY

12.3.1   Faces of convex sets

There are two possible definitions of a face of a (closed) convex set K. Extrinsically, an *exposed face*, or simply *face*, of K is the intersection of K with a support hyperplane. Intrinsically, an *extreme face* of K is a convex subset F of K such that, if $x, y \in K$ and $]xy[\cap F \neq \phi$, then $[xy] \subseteq F$. Faces of each type are closed and convex; exposed faces are extreme, but the converse is generally false. A face F is a *j-face* if $\dim F = j$. Conventionally, we take K to have one (-1)-face ϕ, and, if $\dim K = d$, one d-face K. The set of extreme (exposed) j-faces of K is denoted $\mathscr{E}^j(K)$ ($\mathscr{F}^j(K)$), and the sets

$$\bigcup_{j=-1}^{k} \mathscr{E}^j(K)$$

and

$$\bigcup_{j=-1}^{k} \mathscr{F}^j(K)$$

are the *extreme* and *exposed k-skeleta* of K; the first is written $\mathrm{skel}_k K$. The union of all the extreme (exposed) j-faces of K for $j \leqslant k$ is denoted $\mathrm{ext}_k K$ ($\mathrm{exp}_k K$). The respective d-skeleta of K are denoted $\mathscr{E}(K)$ and $\mathscr{F}(K)$; they are both lattices, partially ordered by inclusion.

We write $\mathrm{ext}_0 K = \mathrm{ext} K$ and $\mathrm{exp}_0 K = \mathrm{exp} K$; these are the *extreme* and *exposed points* of K. We note here the Krein–Milman theorem (due to Minkowski[2] in E^d; see Reference 19): if $K \in \mathscr{K}^d$, then $K = \mathrm{conv}(\mathrm{ext} K)$. Further (Strasewicz, Klee; see Reference 19), $\mathrm{ext} K \subseteq \mathrm{cl}(\mathrm{exp} K)$ and $K = \mathrm{cl}\, \mathrm{conv}(\mathrm{exp} K)$.

The closed convex set P is polyhedral if and only if $\mathscr{E}(P)$ is finite, and in this case, extreme and exposed faces coincide; a polytope is just a bounded polyhedral set. For a polyhedral set P, we write $\mathrm{ext} P = \mathrm{vert} P$, the set of *vertices* of P. We call the 1-faces of P its *edges*, and if $\dim P = k$ its $(k-1)$-faces are its *facets*. Two polytopes are *combinatorially isomorphic* if their lattice of faces are isomorphic. An interesting, but very difficult problem is to enumerate the combinatorial types of d-polytopes with v vertices; apart from results obtained using Gale diagrams (cf. subsection 12.3.2), only sporadic results are known (see Reference 19 and Altshuler and Steinberg[38]). Two polytopes are *dual* if their face lattices are anti-isomorphic; polar polytopes are dual.

Particular classes of polytopes are important. A *simplex* is the convex hull of an affinely independent set of points. A polytope all of whose (proper) faces are simplices is *simplicial*. The dual of a simplicial polytope is *simple*; equivalently, each vertex of a simple d-polytope belongs to d facets. A polytope all of whose faces are isomorphic to cubes is a *cubical* polytope. A *centrally symmetric* polytope P has a centre c, such that $-P = P - 2c$; we usually take $c = o$, the zero vector. *Zonotopes* are vector sums of a finite number of line segments; they are clearly centrally symmetric. A convex set which is a limit of zonotopes is called a *zonoid*. Finally, a polytope P is k-*neighbourly* if every k of its vertices are the vertices of a face of P; if P is centrally symmetric, we take the k vertices to be pairwise non-antipodal. If $\dim P = d$ and $k = [\frac{1}{2}d]$, we simply call P *neighbourly*; examples of neighbourly polytopes are the cyclic polytopes. For all the above, see References 19 and 20.

A d-polytope P is *facet forming* if there is a $(d + 1)$-polytope, all of whose facets are combinatorially isomorphic to P; otherwise P is a *non-facet*. Perles and Shephard[39] demonstrated the existence of a number of non-facets; for example, the d-cross-polytope (analogue of the octahedron) if $d \geqslant 6$.

The 1-skeleton or graph $\mathscr{G}(P)$ of a d-polytope P is d-connected; that is, there are d independent paths in $\mathscr{G}(P)$ between any two vertices of P (Balinski; see Reference 19). More generally, there are d independent paths in $\mathscr{G}(P)$, along which a given linear functional strictly increases from its minimal to its maximal value on P. Further, $\mathscr{F}(P)$, regarded as a complex, is a refinement of $\mathscr{F}(T)$ for a d-simplex T. The only non-trivial result characterising face lattices of polytopes is due to Steinitz[9] (see also Reference 19): a graph is isomorphic to the graph of a 3-polytope if and only if it is planar and 3-connected. A useful technique for investigating polytopes is that of Schlegel diagrams; a *Schlegel diagram* of P is obtained by radial projection from a point just above one facet F of P of the remaining facets of P into F. However, it has been shown that there are complexes, superficially like Schlegel diagrams of 4-polytopes, which are not even isomorphic to Schlegel diagrams. For further discussion in this area, see Grünbaum[19].

Returning to the graph of a d-polytope P, Larman and Mani[40] showed that if any $[\frac{1}{3}(d + 1)]$ pairs of vertices of P are given, there are disjoint edge paths in $\mathscr{G}(P)$ joining corresponding pairs. They conjectured that it should be possible to choose any $[\frac{1}{2}d]$ such pairs, but Gallivan[41] has shown that at most $[\frac{1}{5}(2d + 3)]$ are possible; the exact bound is unknown. For simplicial polytopes, however, Larman and Mani[40] showed that even $[\frac{1}{2}(d + 1)]$ pairs can be given.

In connection with the edge paths, it has long been conjectured that between every two vertices of a d-polytope with n facets is a path of at most $n - d$ edges; this is the *bounded Hirsch conjecture*. The case $n = 2d$ is the most important; this is the *d-step conjecture*. The bounded

Hirsch conjecture is true for $d \leqslant 3$, and the d-step conjecture for $d \leqslant 5$. The corresponding conjectures for unbounded polyhedral sets are known to be false in general. For details, consult Reference 19 or the important paper Klee and Walkup[42]; see also Barnette[43].

We can seek to generalise some of the above ideas to an arbitrary convex body K. It is generally more useful to study the extreme skeleta $\text{skel}_k K$. Then $\text{skel}_1 K$ is connected, and, indeed, d-connected (Larman and Rogers[44]). But there need not be d independent increasing paths, as we had for polytopes; there need not even be a strictly increasing path in $\text{skel}_1 K$ away from a given extreme point (Gallivan[41]). However, there are always at least two strictly increasing paths in the exposed 1-skeleton from bottom to top (Larman and Rogers[45]).

Along slightly different lines, Ewald, Larman and Rogers[46] showed that the directions of line segments in $\text{bd}K$ form a set of σ-finite $(d - 2)$-dimensional Hausdorff measure on S^{d-1}; they give generalisations to r-balls in $\text{bd}K$. A result of Anderson and Klee[47], which is, in a vague sense, dual, is that the set of points in $\text{bd}K$ lying in two support hyperplanes has σ-finite $(d - 2)$-measure.

12.3.2 Diagram techniques

One of the few strikingly new theories of convexity of recent years is that of the various kinds of diagram techniques. The origins of the theory can (in retrospect) be traced back to Whitney[48], in which is founded the theory of matroids, but it first appears in a recognisable form in Gale[49]. The present development of the theory owes most to Perles, whose work is described in Grünbaum[19].

The basic idea is to represent a d-polytope P with n vertices by a certain set of n points in E^{n-d-1}, in one to one correspondence with the vertices of P, called a Gale diagram of P. The original description of these was algebraic (References 19, 20; see also Grünbaum and Shephard[24]; Grünbaum[25]), and would take up too much space here. A geometric formulation due to McMullen (described in Grünbaum and Shephard[24]) is as follows. (A related geometric formulation is given in McMullen and Shephard[50].) An (ordered) set X of n points with $\text{aff}X = E^d$ is affinely equivalent to the image under orthogonal projection of the set of vertices of a regular $(n - 1)$-simplex T in E^{n-1} with centroid at o. (We regard E^d as a coordinate subspace of E^{n-1}.) An (ordered) set $\overline{X}$ which is linearly equivalent to the image of $\text{vert}T$ under orthogonal projection on to the orthogonal complement E^{n-d-1} of E^d is called an *affine* or *Gale transform* of X. We must take the sets to be ordered, since we may have coincidences under the projection.

The fundamental combinatorial correspondence between X and $\bar{X}$ is the *Gale diagram relation.* If $Y \subseteq X$, we write $\bar{Y} = \{\bar{x} \in \bar{X} \mid x \notin Y\}$. Then $F = \text{conv}Y$ is a face of $\text{conv}X$, with $F \cap X = Y$, if and only if $o \in \text{relint conv}\bar{Y}$. This relation clearly gives rise to an *isomorphism* between affine transforms. Any set isomorphic to an affine transform of the set of vertices of a polytope P is called a *Gale diagram* of P.

One useful feature of Gale diagrams is that, if n is not too much larger than d (say $n \leqslant d + 3$ or $d + 4$), then the Gale diagram is small dimensional, and its investigation is intuitively more easy than that of the original polytope. For example, Perles (see Reference 19) has enumerated the combinatorial types of simplicial d-polytopes with $d + 3$ vertices, while Lloyd[51] has solved the same problem for all d-polytopes with $d + 3$ vertices. (See also Altshuler and McMullen[52] and McMullen[53].)

Other applications of Gale diagrams, also due to Perles, are to *projectively unique polytopes,* which are such that any combinatorially isomorphic polytope is actually projectively equivalent (see Grünbaum[19], Perles and Shephard[54] and McMullen[55]), and to *irrational polytopes,* which are such that no combinatorially isomorphic polytope can have all its vertices at points with rational Cartesian coordinates (see Reference 19).

A categorical approach to Gale diagrams is due to Ewald and Voss[56]. A completely different approach is expounded in McMullen[57] (a related, but less versatile, idea is due to Shephard[58]). Let $\mathscr{P}(U)$ denote the class of polyhedral sets of the form

$$P(y) = \{x \in E^d \mid \langle x, u_i \rangle \leqslant \eta_i \; (i = 1, \ldots, n)\}$$

where $y = (\eta_1, \ldots, \eta_n) \in E^n$ and $U = (u_1, \ldots, u_n)$ is a fixed ordered set of vectors spanning E^d. We obtain a *representation* of $\mathscr{P}(U)$ by identifying translates. Since $P(y) + t = P(y')$, where $y' = (\eta'_1, \ldots, \eta'_n)$ and $\eta'_i = \eta_i + \langle t, u_i \rangle \; (i = 1, \ldots, n)$, we do this by taking the image of y under a linear map $\sigma : E^n \to E^{n-d}$ whose kernel consists of the vectors $(\langle t, u_1 \rangle, \ldots, \langle t, u_n \rangle)$, with $t \in E^d$. The position of $p(y) = \sigma(y)$ relative to the vectors $\bar{u}_i = \sigma(e_i)$, with $\{e_1, \ldots, e_n\}$ the standard basis of E^n, determines the combinatorial structure of $P(y)$. There is a connection with Gale diagrams; if $P(y)$ is a polytope, then the image of $\bar{U} = (\bar{u}_1, \ldots, \bar{u}_n)$ under a linear map whose kernel is the line through $p(y)$ is a Gale diagram of the dual of $P(y)$.

There are both metrical and combinatorial applications of this representation technique. For example, a d-polytope with n facets is the Minkowski sum of at most $n - d$ indecomposable polytopes, which admit no non-trivial representation as a sum (McMullen[57]). Representations have also been used to investigate the connection between the following two problems. First, to what extent is the combinatorial type of a polytope determined by the outer normals to its facets? Second, what is the relationship between a polytope and its intersections with its translates? (See McMullen, Schneider and Shephard[59]; compare also Rogers and Shephard[60] and Schneider[61].)

Except for the easy case of the d-cross-polytope, a centrally symmetric d-polytope has at least $2d + 2$ vertices. Thus its Gale diagram has larger dimension than the original polytope, and so has limited usefulness. But there is a suitable variant technique due to Shephard (see McMullen and Shephard[50]); again we give a simplified description. A centrally symmetric set X of $2n$ points spanning E^d is linearly equivalent to the image under orthogonal projection of the set of vertices of a regular n-cross-polytope C in E^n. A set $\overline{X}$ which is linearly equivalent to the image of $\text{vert}\,C$ under orthogonal projection on to the orthogonal complement E^{n-d} of E^d is a *central (c.s.) transform* of X. The combinatorial relationship between X and $\overline{X}$ is more complicated than that for affine transforms, but we still obtain an isomorphism relation, and corresponding *central diagrams* for centrally symmetric polytopes.

In spite of these complications, McMullen and Shephard showed that, if $k(d, m)$ denotes the maximum neighbourliness of a centrally symmetric d-polytope with $2(d + m)$ vertices, then $k(d, 1) = [\frac{1}{2}d]$ and $k(d, 2) = [\frac{1}{3}(d + 1)]$. Schneider[62] has shown that

$$\liminf_{d \to \infty} k(d, m)/(d + m) \geqslant b(m),$$

where $b(m)$ decreases as m increases, and

$$\lim_{m \to \infty} b(m) = 1 - \text{erf}(\sqrt{\log 2}) = 0.2390 \ldots$$

If a polytope P has a certain group of affine symmetries, then a Gale diagram of P can be chosen to have an isomorphic group of linear symmetries, except that we must allow intrinsic symmetries, which permute coincident points of the diagram. Using this idea, it has been possible to enumerate certain combinatorial types of polytopes with symmetries (McMullen and Shephard[63], Ewald-Voss[56]).

Now let $Z = S_1 + \ldots + S_n$, $S_i = [(-x_i)x_i]$ $(i = 1, \ldots, n)$ be a d-zonotope in E^d. There is a combinatorial relationship between Z and a central transform $\overline{X}$ of $X = (\pm x_1, \ldots, \pm x_n)$, which enables us to define a *zonal diagram* of Z. McMullen[64] used this technique to enumerate the combinatorial types of d-zonotopes which are the sum of at most $d + 2$ segments.

If $\overline{X} = (\pm \overline{x}_1, \ldots, \pm \overline{x}_n)$ is a zonal diagram of Z, and we write $\overline{S}_i = [(-\overline{x}_i)\overline{x}_i]$ $(i = 1, \ldots, n)$, then the $(n - d)$-zonotope $\overline{Z} = \overline{S}_1 + \ldots + \overline{S}_i$ is said to be *associated* with Z. McMullen[64] showed that this association is combinatorial in nature, and remarked that it leads to a combinatorial correspondence between arrangements of n hyperplanes in projective space of k and $n - k - 2$ dimensions. Shephard[65] further exploited this association, showing, for example, that if Z and $\bar{Z}$ are dissected (cf. subsection 12.4.1) into cubes with translates of the S_i and $\bar{S}_i$ as sides, then the numbers of cubes in each dissection are the same. Next Shephard[66] investigated zonotopes which tile space; he proved the equivalence of various conditions in small dimensions, of which the most important is that if Z tiles E^d, then $\bar{Z}$ tiles E^{n-d}. The general proof was provided by McMullen[67].

Finally, if X is a positive basis of E^d, then a subset $\bar{X}$ of E_n^{n-d-1} whose affine transform is isomorphic to X is called a *positive diagram* of X. Shephard[68], to whom this concept is due, showed that each vertex of conv$\bar{X}$ is a point of $\bar{X}$ of multiplicity at least 2. Using positive diagrams, he was able to give concise proofs of many of the standard results about positive bases; for example, the theorem of Steinitz that X has at most $2d$ points is a consequence of the remark about conv$\bar{X}$ above.

12.3.3 The numbers of faces of polytopes

If $f_j(P)$ is the number of j-faces of the d-polytope P, the *f-vector* of P is $(f_0(P), \ldots, f_{d-1}(P))$. The basic problem considered in this section is that of classifying the f-vectors of all d-polytopes, or of d-polytopes of certain special kinds. In general, this problem is very far from solution. We first note that the f-vectors satisfy Euler's relation:

$$\sum_{j=0}^{d-1} (-1)^j f_j(P) = 1 - (-1)^d$$

For $d = 3$, the result is due to Euler himself. For $d \geqslant 4$, the relation was discovered by Schläfli[4], although not published at the time. Many incomplete proofs were published between 1880 and 1890; the gap in these proofs was plugged by Bruggesser and Mani[69] only recently, when they showed that the boundary complex of a polytope P can be *shelled* – that is, the facets can be so ordered $F_1, \ldots, F_k$, so that for $r = 1, \ldots,$ $k - 1$, $G_r = F_1 \cup \ldots \cup F_r$ is a topological $(d - 1)$-ball, and $G_r \cap F_{r+1}$ is a topological $(d - 2)$-ball. The first rigorous proof was by Poincaré; simpler proofs are due to Hadwiger, Klee and Grünbaum (see References 19, 20), the last being completely elementary. It can be shown that Euler's relation is essentially the only affine relation satisfied by the f-vectors of all d-polytopes. See Reference 19 for a detailed historical survey of Euler's relation.

But Euler's relation is clearly not the only restriction. For example, Steinitz (see Reference 19) showed that the f-vectors of 3-polytopes are exactly those vectors (f_0, f_1, f_2) satisfying Euler's relation and

$$4 \leqslant f_0 \leqslant 2f_2 - 4, \quad 4 \leqslant f_2 \leqslant 2f_0 - 4$$

However, even for $d = 4$, such a complete classification is unknown; partial results can be found in Reference 19 and Barnette and Reay[70], describing the possible pairs $(f_i(P), f_j(P))$ when $\dim P = 4$.

So, we are led to lower our sights, and ask, for example, for the maximum or minimum of $f_j(P)$ when $f_i(P)$ is fixed. One of these problems is completely solved. Let $f_j(v, d)$ be the number of j-faces of a simplicial neighbourly d-polytope with v vertices. Then we have the upper bound theorem: if P is a d-polytope with v vertices, then $f_j(P) \leqslant f_j(v, d)$, with equality for any $j \geqslant [\frac{1}{2}(d - 1)]$ characterising P as simplicial and neighbourly. This result is due to McMullen (see Reference 20); for earlier partial results, see References 19, 20, 24 and 25.

The proof of the upper bound theorem leads us to restrict our attention to simplicial polytopes. The numbers of faces of these satisfy relations in addition to that of Euler, called the Dehn–Sommerville equations. These have several formulations, but perhaps the most useful are due to McMullen and Walkup[71] (see also Reference 20). For $e \geqslant d$, we write

$$g_k^{(e)} = \sum_{j=-1}^{k} (-1)^{k-j} \binom{e-j-1}{e-k-1} f_j$$

where $f_j = f_j(P)$, and $f_{-1} = 1$. Then we have $g_k^{(e)} = (-1)^{e-d} g_{e-k-2}^{(e)}$ for each k. (The case $e = d$ is due to Sommerville; the numbers $g_k^{(d)}$ are used in the proof of the upper bound theorem.)

There is a recent conjecture about the possible f-vectors of simplicial d-polytopes. If a is a positive integer, the k-canonical expression of a is

$$a = \binom{a_k}{k} + \binom{a_{k-1}}{k-1} + \ldots + \binom{a_i}{i}$$

where the a_j are uniquely defined by $a_k > a_{k-1} > \ldots > a_i \geqslant i$. We write

$$a^{\langle j|k \rangle} = \binom{a_k + j - k}{j} + \binom{a_{k-1} + j - k}{j-1} + \ldots + \binom{a_i + j - k}{i + j - k}$$

omitting binomial coefficients with negative entries. We also write $0^{\langle j|k\rangle} = 0$. Then the so-called g-conjecture of McMullen[72] is: $(f_0, \ldots, f_{d-1})$ is the f-vector of some simplicial d-polytope if and only if $g_k^{(d+1)} = -g_{d-k-1}^{(d+1)}$ for each k, and $g_{k-1}^{(d+1)} (g_k^{(d+1)})^{\langle k|k-1\rangle}$ whenever $g_k^{(d+1)} \geqslant 0$.

This g-conjecture would have the implication $g_k^{(d+1)} \geqslant 0$ for $k = 0$, $\ldots$, $[\tfrac{1}{2}d] - 1$; these inequalities constitute the generalised lower bound conjecture of McMullen and Walkup[71]. The case $k = 0$ is trivial, and the case $k = 1$ is equivalent to the lower bound theorem for simplicial polytopes, proved by Barnette[73]. The g-conjecture itself has been proved for $d \leqslant 5$ or $f_0 \leqslant d + 3$ (McMullen[72]). Finally, by algebraic methods using few properties of polytopes, Stanley[74] has proved that $g_k^{(d)} \leqslant (g_{k-1}^{(d)})^{\langle k+1 k\rangle}$ a result which establishes a stronger form of the upper bound theorem, and which indicates the plausibility of the g-conjecture.

As far as the lower bound for $f_j(P)$ for general d-polytopes P with a given number v of vertices is concerned, there is a conjecture for $v \leqslant 2d$, which is proved for $v \leqslant d + 4$, and a complicated conjecture in case $j = d - 1$, only established for a very restricted range of values of v (McMullen[75]).

The numbers of faces of cubical polytopes satisfy relations analogous to the Dehn–Sommerville equations (see Reference 19), but little is known about the possible f-vectors of these polytopes.

Little is also known about the f-vectors of centrally symmetric, or even centrally symmetric simplicial polytopes. There is even no conjecture about the maximum possible numbers of faces; the situation is complicated here by the fact that a centrally symmetric d-polytope with too many vertices cannot be neighbourly (cf. subsection 12.3.2).

Much more is known about 3-polytopes, and particularly about simple 3-polytopes. If $p_j = p_j(P)$ is the number of j-gonal faces of a simple 3-polytope P, then

$$3p_3 + 2p_4 + p_5 = 12 + \sum_{j \geqslant 7} (j-6)p_j$$

(see Reference 19). Eberhard[5] showed that, given a sequence $(p_3, p_4, p_5, p_7, \ldots)$ satisfying this equation, there is a p_6 and a simple 3-polytope P with $p_j(P) = p_j$ for all j. Indeed, if $p_3 = p_4 = 0$, we can take any $p_6 \geqslant 8$ (Grünbaum[76]). If p_3 or p_4 is non-zero, there are other restrictions on p_6; for a survey of these and related results, see Grünbaum[25]. Analogous results hold for 3-polytopes with 4-valent vertices.

12.4 METRICAL THEORY

12.4.1 Valuations and volume

In developing the theory of volume of convex sets, we could just assume the theory of Lebesgue measure. However, there is an alternative approach, basically due to Hadwiger[12], which leads to many interesting and fruitful generalisations.

A valuation is a function φ on $\mathscr{P}^d$ or $\mathscr{K}^d$, taking values in some real linear space, such that $\varphi(K_1 \cup K_2) + \varphi(K_1 \cap K_2) = \varphi(K_1) + \varphi(K_2)$, whenever $K_1 \cup K_2$ is convex. We say φ is simple if $\varphi(K) = 0$ when $\dim K < d$. If $\mathscr{G}$ is a group of Euclidean motions, we say φ is a $\mathscr{G}$-valuation if $\varphi(\Psi K) = \varphi K$ for all $\Psi \in \mathscr{G}$.

Hadwiger[12] characterises volume on $\mathscr{P}^d$ as the unique non-negative (real-valued) translation invariant simple valuation which takes the value 1 on the standard unit cube. In the course of his proof, he shows that a translation invariant simple valuation ϕ admits a polynomial expansion

$$\varphi(\lambda P) = \sum_{r=1}^{d} \lambda^r \varphi_r(P)$$

for non-negative rational λ. (If φ is continuous or monotone, λ can be real.) For volume V, non-negativity implies monotoneity, and this enables one to define the Jordan volume of a general $K \in \mathscr{K}^d$ by

$$V(K) = \sup\{V(P)|P \in \mathscr{P}^d, P \subseteq K\} = \inf\{V(Q)|Q \in \mathscr{P}^d, K \subseteq Q\}$$

In fact, one can use the inclusion–exclusion relation, which follows from the valuation property (Volland[77], Sallee[78]), to define the volume of any finite union of compact convex sets. In a different direction, one can also obtain Lebesgue measure. One particular consequence of Hadwiger's approach is that volume is rigid motion invariant.

An idea at the basis of the above discussion is that of dissections. An expression $P = P_1 \cup \ldots \cup P_k$, where $\dim(P_i \cap P_j) < d$ if $i \neq j$, is a dissection of the d-polytope P. Let $\mathscr{G}$ be a group of motions, as above. If $Q \in \mathscr{P}^d$ has a dissection $Q = Q_1 \cup \ldots \cup Q_k$, such that $Q_i = \Psi_i P_i$ for some $\Psi_i \in \mathscr{G}$ $(i = 1, \ldots, k)$, we call P and Q $\mathscr{G}$-equidissectable. Recently, necessary and sufficient conditions for two polytopes to be equidissectable by translation, conjectured by Hadwiger[79] (see also Reference 12), have been proved by Jessen and Thorup[80]; these conditions involve certain classes of simple valuations. There are conjectured conditions for equidissectability by rigid motions (see Reference 12), but this problem remains open for $d \geqslant 4$.

To discuss general valuations, we need some preliminary material. The angle of a convex cone C with apex a is that proportion of a ball in aff*C* with centre *a* which lies in *C* (thus all angles are normalised and measured intrinsically). If *F* is a face of a polyhedral set *P*, we define the *internal angle* β(*F, P*) of *P* at *F* to be the angle of the cone generated by *P* with a point of relint*F* as apex, and the *external angle* γ(*F, P*) of *P* at *F* to be the angle of the cone of outer normal vectors to support hyperplanes of *P* which contain *F*. These internal and external angles satisfy various quadratic relations, which lead to the inversion formulae: if φ and ψ are functions on polytopes, then the following are equivalent:

$$\psi(P) = \Sigma_F (-1)^{\dim P - \dim F} \beta(F, P) \varphi(F)$$

$$\varphi(P) = \Sigma_F \gamma(F, P) \psi(F)$$

where the sums extend over all non-empty faces *F* of *P* (including *P* itself). This is due to McMullen[81]. Now, if φ is a valuation, then ψ is (essentially) a simple valuation (McMullen[82]).

We may now deduce that general translation invariant valuations have polynomial expansions

$$\varphi(\lambda P) = \sum_{r=0}^{d} \lambda^r \varphi_r(P);$$

if φ is continuous, the coefficients can be non-negative real numbers, and the arguments any compact convex sets. But we even have polynomial expansions for φ(λ₁*P*₁ + . . . + λ_k*P_k*). The coefficients we call *mixed valuations*; each is a homogeneous translation invariant valuation of appropriate degree in its various arguments. (For all this, see McMullen[82].) When the original valuation is volume, we obtain the well-known *mixed volumes* (see References 1, 11, 16); the most general mixed volume is *V*(*K*₁, . . ., *K_d*), where *d* ! *V*(*K*₁, . . . *K_d*) is the coefficient of λ₁ . . . λ_d in *V*(λ₁*K*₁ + λ_d*K_d*).

As a particularly important example, we write

$$V(K + \rho B^d) = \sum_{i=0}^{d} \binom{d}{i} \rho^i W_i(K),$$

where W_i (which is homogeneous of degree $d - i$) is the *i*-th *quermassintegral* (see References 1, 8, 12; cf. subsection 12.4.4 for alternative definitions). Hadwiger[12] has characterised (non-negative) linear combinations of quermassintegrals as the continuous (monotone) rigid motion invariant valuations.

We can extend the scope of this theory to *translation equivariant* valuations, which under translation obey a rule of the form φ(*P* + *t*) = φ(*P*) + ϑ(*P, t*), where ϑ is linear in *t*. In particular, starting from the moment vector (Zielvektor) *z*(*K*) = *V*(*K*)*g*(*K*), where *g*(*K*) is the centre of gravity of K, Schneider[83] has defined mixed moment vectors, and Hadwiger and Schneider[84] have characterised the quermass-vectors, which are vectorial analogues of the quermassintegrals.

If φ is a valuation, and we define φ^* by

$$\varphi^*(P) = \sum_F (-1)^{\dim F} \varphi(F)$$

then φ^* is also a valuation (Sallee[78], McMullen[82]). We say that φ satisfies an Euler-type relation if $\varphi^*(P) = \pm\varphi(\pm P)$ (for some fixed choice of signs). Shephard[85] found many examples of Euler-type relations, including ones involving mixed volumes (see also Reference 19). McMullen[82] has shown that these are cases of a general theorem: if φ is a homogeneous translation equivariant valuation of degree r, then $\varphi^*(P) = (-1)^r \varphi(-P)$ for each $P \in \mathscr{P}^d$.

There are Euler-type theorems involving angles alone, such as Gram's theorem:

$$\sum_F (-1)^{\dim F} \beta(F, P) = 0$$

(if $\dim P > 0$), theorems on angle deficiencies (due to Perles and Shephard) and Grassmann angles (due to Grünbaum). These are all special cases of general relations (see McMullen[81] for details and references).

Somewhat related ideas are due to Schneider. An *endomorphism* of $\mathscr{K}^d$ is a map $\Phi : \mathscr{K}^d \to \mathscr{K}^d$ such that $\Phi(K_1 + K_2) = \Phi(K_1) + \Phi(K_2)$. If Φ is continuous and volume preserving, then there is a volume preserving affinity Ψ of E^d such that $\Phi(K)$ is a translate of $\Psi(K)$ for all $K \in \mathscr{K}^d$ (Schneider[86]). Schneider[87] also characterises continuous endomorphisms which commute with rigid motions. In a slightly different vein, Schneider[88] shows that an isometry of $\mathscr{K}^d$ is induced by an isometry of E^d.

12.4.2 Isoperimetric theorems

The discussion of this section mainly centres around inequalities between mixed volumes. Our starting point is the Brunn–Minkowski theorem, which states that the dth root of volume is concave. Moreover, if

$$V^{1/d}((1-\lambda)K_0 + \lambda K_1) = (1-\lambda) V^{1/d}(K_0) + \lambda V^{1/d}(K_1)$$

for some $0 < \lambda < 1$, then K_0 and K_1 are homothetic ($K_1 = \mu K_0 + t$, for some $\mu > 0$ and $t \in E^d$), or one is a point, or they lie in parallel hyperplanes. Four different proofs are known; see References 1, 11, 12. Grünbaum[19] has historical information.

Let us write

$$V(\lambda_0 K_0 + \lambda_1 K_1) = \sum_{i=0}^{d} \binom{d}{i} \lambda_0^{d-i} \lambda_1^i V_i$$

Then by considering the first derivative of

$$f(\lambda) = V^{1/d}((1-\lambda)K_0 + \lambda K_1) - (1-\lambda)V_0^{1/d} - \lambda V_d^{1/d}$$

at 0 we obtain Minkowski's first inequality: $V_1^d \geqslant V_0^{d-1} V_d$. Moreover, if V_0, $V_d > 0$, we have equality only if K_0 and K_1 are homothetic. Considering the second derivative of f at 0 yields Minkowski's second inequality: $V_1^2 \geqslant V_0 V_2$; equality can occur if K_0 and K_1 are not homothetic (see Reference 1).

The second Minkowski inequality is the starting point for proving the Aleksandrov–Fenchel inequalities: if $K_1, \ldots, K_d \in \mathscr{K}^d$, then

$$V(K_1, K_2, K_3, \ldots, K_d)^2 \geqslant V(K_1, K_1, K_3, \ldots, K_d)V(K_2, K_2, K_3, \ldots, K_d)$$

(See Aleksandrov[89] and Busemann[16]; the latter proof can be shortened using Gårding[90] or Schneider[91].) From the Aleksandrov–Fenchel inequalities can be deduced many others; Shephard[92] discusses these in some detail. See also Busemann[16] and Hadwiger[12].

The mixed volume $dV(K, \ldots, K, B^d) = S(K)$ is just the *surface area* of K. A particular case of the first Minkowski inequality is thus the classical isoperimetric inequality: $V(K)^{d-1} \leqslant (d^{-1}S(K))^d / \kappa_d$, where $\kappa_d = V(B^d)$. Equality (if $S(K) > 0$) occurs only if K is a ball.

If we reverse the roles of K and B^d we similarly obtain $V(K) \leqslant W_{d-1}(K)^d / \kappa_d^{d-1}$. Now $W_{d-1}(K) = \frac{1}{2}\kappa_d b(K)$, where

$$b(K) = (d\kappa_d)^{-1} \int_{S^{d-1}} \{h(K, u) + h(K, -u)\} \, du$$

is the *mean width* of K (compare subsection 12.4.5 below). Since $b(K) \leqslant D(K)$, the diameter of K, we also obtain $V(K) \leqslant \kappa_d(\frac{1}{2}D(K))^d$. Equality in both relations occurs only for balls.

There are many generalisations and stronger versions of the isoperimetric theorems. For example, we have (Hadwiger[12])

$$(d^{-1}S(K))^d - \kappa_d V(K)^{d-1} \geqslant [(d^{-1}S(K))^{1/(d-1)} - \kappa_d^{1/(d-1)} r(K)]^{d(d-1)}$$

where $r(K)$ is the *inradius* of K, the radius of the largest ball contained in K. (See also subsection 12.4.4.)

There are also isoperimetric theorems for various classes of convex bodies, especially for polytopes. Fejes Tóth[22] characterises many regular polytopes by means of extremal properties. The study of isoperimetrically best representatives of a given combinatorial type leads to Lindelöf's theorem (see Reference 12): among all d-polytopes with a given volume whose facets have given outer normal vectors, the smallest surface area occurs just when the facets touch some sphere. Steinitz (see Reference 19) showed that there are some combinatorial types of 3-polytopes containing none all of whose facets touch a sphere; these have no isoperimetrically best representatives.

12.3.3 The analytic theory

So far, we have not paid too much specific attention to the analytic theory of convex sets, even though in many cases it has historical precedence. There are two ways of approaching this theory. The first is to regard a convex surface as, locally at least, the graph of a convex function; we have implicitly touched on this in subsection 12.22. The other is to regard a convex surface as the envelope of its support hyperplanes. Thus we shall first look more closely at support properties and the support function.

The support function of the face $F(K, u)$ of $K \in \mathcal{H}^d$ in direction u is the directional derivative $h'(K, u; .)$ of the support function of K. Hence, if $h(K, .)$ is differentiable at u, $F(K, u)$ is the single point $\nabla h(K, u)$. As we remarked in subsection 12.22, the convex function $h(K, u)$ is twice differentiable almost everywhere. If $h(K, .)$ is twice differentiable at u, then one eigenvalue of the matrix with entries the second derivatives h_{ij} is 0; the others are the principal radii of curvature $R_1, \ldots, R_{d-1}$ of bd$K$ at $\nabla h(K, u)$. Thus, if $d\omega(u)$ is the element of area on S^{d-1} at u, the corresponding element of area on bd$K$ is $R_1 \ldots R_{d-1}\, d\omega(u) = D_{d-1}h\,d\omega(u)$, where we write $D_r h$ for the rth symmetric function of $R_1, \ldots, R_{d-1}$, and so the sum of the principal $r \times r$ minors of (h_{ij}) (see References 1, 16).

Now let us assume that $h(K, .)$ (or, more briefly, K) is analytic. We then have two expressions for the quermassintegral $W_i(K)$:

$$W_i(K) = \left\{ d\binom{d-1}{d-i} \right\}^{-1} \int_{S^{d-1}} D_{d-i}h\,d\omega(u)$$

$$= \left\{ d\binom{d-1}{d-i-1} \right\}^{-1} \int_{S^{d-1}} h(K, u)D_{d-i-1}h\,d\omega(u)$$

Recall that $W_0(K) = V(K)$ and $dW_1(K) = S(K)$. More generally, since

$$h(\lambda_1 K_1 + \ldots + \lambda_k K_k, u) = \lambda_1 h(K_1, u) + \ldots + \lambda_k h(K_k, u)$$

we can easily construct an analytic theory of mixed volumes. It is then possible to prove the Brunn–Minkowski theorem (see Reference 1) and the Aleksandrov–Fenchel inequalities (Aleksandrov[89]; see also Busemann[16]) by analytic methods.

If $h(K, .)$ is not everywhere twice differentiable, we can adopt a different approach. Let ω be a Borel set on the unit sphere S^{d-1} and A its inverse spherical image on bd$K$. Then the area $S(K; \omega)$ of A has the expansion

$$S(K + \lambda B^d; \omega) = \sum_{k=0}^{d-1} \binom{d-1}{k} \lambda^{d-k-1} S_k(K; \omega)$$

for $\lambda \geq 0$. The function $S_k(K; .)$ is the k-th area measure (Fenchel and Jessen[93]); $S_k(K; d\omega(u))$ replaces $D_{d-k-1} h d\omega(u)$. For an alternative definition, see Schneider[94]. Federer[95] has introduced a different, but closely related, area measure.

Because of the translation invariance, the surface area measures $S_k(K; .)$ satisfy $\int u S_k(K; d\omega(u)) = o$. If S is a non-negative measure on S^{d-1}, whose support does not lie in any hemisphere, then $S = S_{d-1}(K; .)$ for some $K \in \mathscr{H}^d$, which is unique up to translation (Aleksandrov[89], Fenchel and Jessen[93]; the result for polytopes, and the uniqueness, are due to Minkowski[2]; see also Reference 1).

This result enables us to define a new addition for convex bodies, denoted #, and called Blaschke addition. It is defined by

$$S_{d-1}(K_1 \# K_2; .) = S_{d-1}(K_1; .) + S_{d-1}(K_2; .)$$

(Blaschke[7], Grünbaum[19], Firey and Grünbaum[96]). Firey[97] has discussed several connections between Blaschke and Minkowski addition in three dimensions; in E^2 they coincide.

We cannot end this section without mentioning the subject of rigidity of convex surfaces, which has received much attention from Russian workers in convexity (Aleksandrov[15,21], Lyusternik[23], Efimov[98], Pogorelov[99 -101]; see also Busemann[16]). A convex surface has an intrinsic metric, induced by its geodesics. The basic questions are whether a convex surface admits two non-congruent embeddings, or can be deformed, while remaining a convex surface with the same intrinsic metric. For the boundaries of 3-polytopes, Cauchy (see Lyusternik[23]) already knew the answer to be negative; the same result has now been established for many other kinds of convex surface, some not necessarily closed. A closely related question concerns the existence of a convex surface realising a given intrinsic metric.

12.4.4   Integral geometry

The starting point for the discussion of this section has little to do with convexity. It is the fact that the rotation group and the group of rigid motions possess invariant measures, the Haar measures, which are unique except for a normalisation factor. There are induced measures on the families of linear and affine r-flats, for each r. These measures can be expressed as the integrals of differential forms in suitable coordinate systems. For an introduction to this theory, where the applications to convexity are borne in mind, see Santaló[18].

We shall write L^r and A^r for typical linear and affine r-flats, and dL^r and dA^r for the corresponding differential forms. One fact to note first is that if $\mathscr{M}$ is a subset of the Grassmannian $\mathscr{G}^{d, r}$ of all linear r-flats, and

$$\bar{\mathscr{M}} = \{L^{d-r} \in \mathscr{G}^{.d, d-r} | (L^{d-r})^\perp \in \mathscr{M}\}$$

is the corresponding subset of orthogonal $(d - r)$-flats, then

$$\int_{\mathscr{M}} dL^r = \int_{\bar{\mathscr{M}}} dL^{d-r}.$$

We write Φ and Ψ for typical rotations and rigid motions, respectively.

We first note the integral geometric definitions of the quermass-integrals:

$$W_i(K) = c \int \chi(K \cap A^i) dA^i = c \int V_{d-i}(K|L^{d-i}) dL^{d-i}$$

where χ is the Euler characteristic, $K|L^{d-i}$ is the image of K under orthogonal projection on to L^{d-i}, V_{d-i} denotes $(d-i)$-dimensional volume, and c is a normalisation constant, possible different at each occurrence (Reference 12; see also Matheron[102]). The latter definition gives what is sometimes called the *outer quermassintegral*; if $a \in \text{int}K$, the *inner quermassintegral* of K with respect to a is:

$$\widetilde{W}_i(K-a) = c \int V_{d-i}((K-a) \cap L^{d-i}) dL^{d-i}$$

Blaschke[8] gives alternative expressions for the inner quermassintegrals.

Of interest in this connection is the recent result of Larman and Rogers[103]; there are convex bodies K_1 and K_2 in E^d (for each $d \geqslant 12$, at least), centrally symmetric about o, such that $V_{d-1}(K_1 \cap H) < V_{d-1}(K_2 \cap H)$ for each hyperplane H through o, but with $V(K_1) > V(K_2)$. We can even choose K_2 to be a ball. The proof uses probabilistic arguments. The analogous result for projections was proved earlier by Schneider[104]: if K_1 is any centrally symmetric convex body which is not a zonoid, then there is a centrally symmetric convex body K_2, such that $V_{d-1}(K_1|H) < V_{d-1}(K_2|H)$ for each hyperplane through o, but with $V(K_1) > V(K_2)$.

Of great theoretical importance are the kinematic formulae (Blaschke[8], Hadwiger[12], Matheron[102]), which are

$$c \int W_i(K_0 \cap \Psi K_1) d\Psi = \sum_{j=0}^{i} \binom{i}{j} W_j(K_0) W_{i-j}(K_1)$$

$$c \int W_i(K_0 + \Phi K_1) d\Phi = \sum_{j=0}^{d-i} \binom{d-i}{j} W_j(K_0) W_{d-i-j}(K_1)$$

For applications of these and other kinematic formulae, see Hadwiger[12].

The most extensive study of integral geometry has been in E^2; we refer the reader particularly to Kendall and Moran[105] and Santaló[18]. There are very many elegant and beautiful results here; we just mention two which can be stated easily, both stronger forms of the isoperimetric inequality. Let K be a plane convex body, with area A and perimeter L. First, let M_k be the measure of those rigid motions Ψ of E^2 such that bd$K$ and bdΨK meet in k points. Then

$$L^2 - 4\pi A = M_4 + 2M_6 + 3M_8 + \ldots$$

In particular, if K is not a disc, then there is some Ψ such that the boundaries of K and ΨK meet in at least four points. Second, if r is the inradius and R the circumradius of K, then Bonnesen showed that

$$L^2 - 4\pi A \geqslant \pi^2 (R - r)^2$$

For both these results, see Santaló[18].

12.4.5 Miscellaneous problems

There is a considerable variety of special problems in convexity, and we have no room to mention even a small fraction of them. However, there are a few which we cannot avoid; they can loosely be classified as covering problems.

The well known theorem of Jung states that a subset of E^d of diameter D is contained in some ball of radius $D\sqrt{[d/2(d+1)]}$. If K is a convex body, then there is a unique ellipsoid E of minimal volume, the Löwner ellipsoid, which contains K. If o is the centre of E, then $d^{-1}E \subseteq K$, while if K is centrally symmetric, then $d^{-\frac{1}{2}} E \subseteq K$ (John[106]). In a similar spirit, there is a unique ellipsoid of maximal volume contained in K. For both results, see John[106] or Danzer, Laugwitz and Lenz[107].

Borsuk's famous conjecture is that a set of diameter D in E^d can be covered by at most $d + 1$ sets of diameter less than D. It is enough to prove this for convex sets, and indeed, for sets $K \in \mathscr{K}^d$ of constant width. The *width* of K in direction $u \in S^{d-1}$ is $h(K, u) + h(K, -u)$, the distance between the support hyperplanes of K which are perpendicular to u. The conjecture is known to be true if $d \leqslant 3$, and also if K is smooth. See Grünbaum[108] for a complete discussion of this and related problems. Knast[109] has recently remarked that it is an easy consequence of Jung's theorem that 2^d sets of diameter less than D will cover K.

Another famous problem was proposed by Tarski. A *plank* is a closed convex set bounded by parallel hyperplanes. The original plank problem, solved by Bang[110], states that if a convex set of minimal width δ is covered by planks, then the sum of their widths is at least δ. If P is a plank of width α, and β is the width of K in the same direction, then α/β is the relative width of P. The generalised plank problem (Bang[110]), still unsolved, conjectures that the sum of the relative widths of a covering of K by planks is at least 1.

REFERENCES

1. Bonnesen, T. and Fenchel, W., *Theorie der konvexen Körper,* Springer, Berlin (1934); reprinted Chelsea, New York (1948)
2. Minkowski, H., *Gesammelte Abhandlungen,* 2 vols, Teubner, Berlin and Leipzig (1911)
3. Steiner, J., *Gesammelte Werke,* 2 vols, Reimer, Leipzig (1881, 1882)
4. Schläfli, L., 'Theorie der vielfachen kontinuität' in *Gesammelte Abhandlungen,* Birkhäuser, Basel (1950)
5. Eberhard, V., *Zur Morphologie der Polyeder,* Teubner, Leipzig (1891)
6. Brückner, M., *Vielecke und Vielfache,* Teubner, Leipzig (1900)
7. Blaschke, W., *Kreis und Kugel,* Verlag von Veit, Leipzig (1916); reprinted Chelsea, New York (1949); 2nd edn, Gruyter, Berlin (1956)
8. Blaschke, W., *Integralgeometrie,* Chelsea, New York (1948)
9. Steinitz, E., 'Polyeder und Raumeinteilung' in *Enzyklopädie der mathematischen Wissenschaften,* Vol. 3 (Geometrie), Part 3AB12, Teubner, Leipzig (1922)
10. Steinitz, E. and Rademacher, H., *Vorlesungen über die Theorie der Polyeder,* Springer, Berlin (1934)
11. Eggleston, H.G., *Convexity,* Cambridge University Press, Cambridge (1958)
12. Hadwiger, H., *Vorlesungen über Inhalt, Oberfläche, und Isoperimetrie,* Springer, Berlin (1957)
13. Rockafellar, R.T., *Convex analysis,* Princeton University Press, Princeton (1970)
14. Valentine, F.A., *Convex sets,* McGraw-Hill, New York (1964)
15. Aleksandrov, A.D., *Die innere Geometrie der konvexen Flächen,* Akademie Verlag, Berlin (1955); translated from Russian
16. Busemann, H., *Convex surfaces,* Wiley, New York (1958)
17. Hadwiger, H., *Altes und Neues über konvexe Körper,* Birkhäuser, Basel (1955)

18. Santaló, L.A., *Introduction to integral geometry*, Hermann, Paris (1953)
19. Grünbaum, B., *Convex polytopes*, Wiley, New York (1967)
20. McMullen, P. and Shephard, G.C., *Convex polytopes and the upper bound conjecture*, London Mathematical Society Lecture Note Series 3, Cambridge University Press, Cambridge (1971)
21. Aleksandrov, A.D., *Konvexe Polyeder*, Akademie Verlag, Berlin (1958); translated from Russian
22. Fejes Tóth, L., *Regular figures*, Pergamon Press, New York (1964)
23. Lyusternik, L.A., *Convex figures and polyhedra*, Dover, New York (1963); translated from Russian
24. Grünbaum, B. and Shephard, G.C., 'Convex polytopes', *London Mathematical Society. Bulletin*, **1**, 257-300 (1969)
25. Grünbaum, B., 'Polytopes, graphs and complexes', *American Mathematical Society. Bulletin*, **76**, 1131-1201 (1970)
26. *American Mathematical Society. Proceedings of Symposia in Pure Mathematics*, **7**: (1963)
27. Fenchel, W. (ed.), *Proceedings of the colloquium on convexity, Copenhagen, 1965*, Københavns Universitets Matematiske Institut, Copenhagen (1967)
28. Reay, J.R., 'Generalizations of a theorem of Carathéodory', *American Mathematical Society. Memoirs*, **54** (1965)
29. Tverberg, H., 'A generalization of Radon's theorem', *London Mathematical Society. Journal*, **41**, 123-138 (1966)
30. Reay, J.R., 'An extension of Radon's theorem', *Illinois Journal of Mathematics*, **12**, 184-189 (1968)
31. Danzer, L., Grünbaum, B. and Klee, V.L., 'Helly's theorem and its relatives', *American Mathematical Society. Proceedings of Symposia in Pure Mathematics*, **7**, 101-180 (1963)
32. McMullen, P., 'The support functions of compact convex sets', *Elemente der Mathematik*, **31**, 117-119 (1976)
33. Busemann, H., 'The foundations of Minkowskian geometry', *Commentarii Mathematici Helvetici*, **4**, 156-187 (1950)
34. Hammer, P.C., 'Approximations of convex surfaces by algebraic surfaces', *Mathematika*, **10**, 64-71 (1963)
35. Firey, W.J., 'Approximating convex bodies by algebraic ones', *Archiv der Mathematik*, **25**, 424-425 (1974)
36. Busemann, H. and Feller, W. '*Krümmungseigenschaften konvexer Flächen*', *Acta Mathematica*, **66**, 1-47 (1935)
37. Aleksandrov, A.D., 'Almost everywhere existence of second differentials of convex functions', *Leningradskiĭ Universitet. Uchenye Zapiski (Mathematics Series)*, **6**, 3-35 (1939) [in Russian]
38. Altshuler, A. and Steinberg, L., 'Neighborly 4-polytopes with 9 vertices', *Journal of Combinatorial Theory. A*, **15**, 270-287 (1973)
39. Perles, M.A. and Shephard, G.C., 'Facets and non-facets of convex polytopes', *Acta Mathematica*, **119**, 113-145 (1967)
40. Larman, D.G. and Mani, P., 'On the existence of certain configurations within graphs and the 1-skeletons of polytopes', *London Mathematical Society. Proceedings*, **20**, 144-160 (1970)
41. Gallivan, S., *Properties of the one-skeleton of a convex body*, Ph.D. thesis, University of London (1974)
42. Klee, V.L. and Walkup, D.W., 'The *d*-step conjecture for polyhedra of dimension $d < 6$', *Acta Mathematica*, **117**, 53-78 (1967)
43. Barnette, D.W., 'An upper bound for the diameter of a polytope', *Discrete Mathematics*, **10**, 9-13 (1974)

ISRD–8

44. Larman, D.G. and Rogers, C.A., 'Paths in the one-skeleton of a convex body', *Mathematika*, 17, 293-314 (1970)
45. Larman, D.G. and Rogers, C.A., 'Increasing paths in the one-skeleton of a convex body and the directions of line segments in the boundary of a convex body', *London Mathematical Society. Proceedings*, 23, 683-698 (1971)
46. Ewald, G., Larman, D.G. and Rogers, C.A., 'The directions of the line segments and of the r-dimensional balls on the boundary of a convex body in euclidean space', *Mathematika*, 17, 1-20 (1970)
47. Anderson, R.D. and Klee, V.L., 'Convex functions and upper semi-continuous collections', *Duke Mathematical Journal*, 19, 349-357 (1952)
48. Whitney, H., 'On the abstract properties of linear dependence', *American Journal of Mathematics*, 57, 509-533 (1935)
49. Gale, D., 'Neighboring vertices on a convex polyhedron' in *Linear inequalities and related systems* (ed. H.W. Kuhn and A.W. Tucker), Princeton University Press, Princeton, N.J. (1956)
50. McMullen, P. and Shephard, G.C., 'Diagrams for centrally symmetric polytopes', *Mathematika*, 15, 123-138 (1968)
51. Lloyd, E.K., 'The number of d-polytopes with d + 3 vertices', *Mathematika*, 17, 120-132 (1970)
52. Altshuler, A. and McMullen, P., 'The number of simplicial neighbourly d-polytopes with d + 3 vertices' *Mathematika*, 20, 263-266 (1973)
53. McMullen, P., 'The number of neighbourly d-polytopes with d + 3 vertices', *Mathematika*, 21, 26-31 (1974)
54. Perles, M.A. and Shephard, G.C.,'A construction for projectively unique polytopes', *Geometriae Dedicata*, 3, 357-363 (1974)
55. McMullen, P., 'Constructions for projectively unique polytopes', *Discrete Mathematics*, 14, 347-358 (1976)
56. Ewald, G. and Voss, K., 'Konvexe Polytope mit Symmetriegruppe', *Commentarii Mathematici Helvetici*, 48, 137-150 (1973)
57. McMullen, P., 'Representations of polytopes and polyhedral sets', *Geometriae Dedicata*, 2, 83-99 (1973)
58. Shephard, G.C., 'Polyhedral diagrams for sections of the non-negative orthant', *Mathematika*, 18, 55-263 (1971)
59. McMullen, P., Schneider, R. and Shephard, G.C., 'Monotypic polytopes and their intersection properties', *Geometriae Dedicata*, 3, 99-129 (1974)
60. Rogers, C.A. and Shephard, G.C., 'The difference body of a convex body', *London Mathematical Society. Journal*, 33, 270-281 (1958)
61. Schneider, R., 'Über die Durchschnitte translationsgleicher konvexer Körper und eine Klasse konvexer Polyeder', *Hamburg, Universität: Mathematisches Seminar. Abhandlungen*, 30, 118-128 (1967)
62. Schneider, R., 'Neighbourliness of centrally symmetric polytopes in high dimensions', *Mathematika*, 22, 176-181 (1975)
63. McMullen, P. and Shephard, G.C., 'Polytopes with an axis of symmetry', *Canadian Journal of Mathematics*, 22, 265-287 (1970)
64. McMullen, P., 'On zonotopes', *American Mathematical Society. Transactions*, 159, 91-110 (1971)
65. Shephard, G.C., 'Combinatorial properties of associated zonotopes', *Canadian Journal of Mathematics*, 26, 302-321 (1974)
66. Shephard, G.C., 'Space filling zonotopes', *Mathematika*, 21, 261-269 (1974)
67. McMullen, P., 'Space tiling zonotopes', *Mathematika*, 22, 202-211 (1975)
68. Shephard, G.C., 'Diagrams for positive bases', *London Mathematical Society. Journal*, 4, 165-175 (1971)
69. Bruggesser, H. and Mani, P., 'Shellable decompositions of cells and spheres', *Mathematica Scandinavica*, 17, 179-184 (1972)

70. Barnette, D.W. and Reay, J.R., 'Projections of *f*-vectors of 4-polytopes', *Journal of Combinatorial Theory. A*, **15**, 200-209 (1973)
71. McMullen, P., and Walkup, D.W., 'A generalized lower-bound conjecture for simplicial polytopes', *Mathematika*, **18**, 264-273 (1971)
72. McMullen, P., 'The numbers of faces of a simplicial polytope', *Israel Journal of Mathematics*, **9**, 559-570 (1971)
73. Barnette, D.W., 'A proof of the lower bound conjecture for convex polytopes', *Pacific Journal of Mathematics*, **46**, 349-354 (1973)
74. Stanley, R.P., 'The upper bound conjecture and Cohn–Macauley rings', *Studies in Applied Mathematics*, **54**, 135-142 (1975)
75. McMullen, P., 'The minimum number of facets of a convex polytope', *London Mathematical Society. Journal*, **3**, 350-354 (1971)
76. Grünbaum, B., 'Some analogues of Eberhard's theorem on convex polytopes', *Israel Journal of Mathematics*, **6**, 398-411 (1968)
77. Volland, W., 'Ein Fortsetzungssatz für additive Eipolyederfunktionale im euklidischen Raum', *Archiv der Mathematik*, **8**, 144-149 (1957)
78. Sallee, G.T., 'Polytopes, valuations and the Euler relation', *Canadian Journal of Mathematics*, **20**, 1412-1424 (1968)
79. Hadwiger, H., 'Translative Zerlegungsgleichheit der Polyeder der gewöhnlichen Raum', *Journal für die Reine und Angewandte Mathematik*, **233**, 200-212 (1968)
80. Jessen, B. and Thorup, A., 'The algebra of polytopes in affine spaces', (to appear)
81. McMullen, P., 'Non-linear angle-sum relations for polyhedral cones and polytopes', *Cambridge Philosophical Society. Proceedings*, **78**, 247-261 (1975)
82. McMullen, P., 'Valuations and Euler-type relations on certain classes of convex polytopes' *London Mathematical Society. Proceedings*, **35** (1977) [to appear]
83. Schneider, R., 'Additive Transformationen konvexer Körper', *Geometriae Dedicata*, **3**, 221-228 (1974)
84. Hadwiger, H. and Schneider, R., 'Vektorielle Integralgeometrie', *Elemente der Mathematik*, **26**, 49-57 (1971)
85. Shephard, G.C., 'Euler-type relations for convex polytopes', *London Mathematical Society. Proceedings*, **18**, 597-606 (1968)
86. Schneider, R., 'Additive Transformationen konvexer Körper', *Geometriae Dedicata*, **3**, 221-228 (1974)
87. Schneider, R., 'Equivalent endomorphisms of the space of convex bodies', *American Mathematical Society. Transactions*, **194**, 53-78 (1974)
88. Schneider, R., 'Isometrien des Raumes der konvexen Körper, *Colloquium Mathematicum*, **33**, 219-224, 304 (1975)
89. Aleksandrov, A.D., 'On the theory of mixed volumes', *Matematicheskiĭ Sbornik*, **44**, (Novaya seriya 2), 947-97, 1205-1238 (1937); **45** (Novaya seriya 3), 27-46, 227-251 (1938) [in Russian]
90. Gårding, L., 'An inequality for hyperbolic polynomials', *Journal of Mathematics and Mechanics*, **8**, 957-965 (1959)
91. Schneider, R., 'On Aleksandrov's inequalities for mixed discriminants', *Journal of Mathematics and Mechanics*, **15**, 285-290 (1966)
92. Shephard, G.C., 'Inequalities between mixed volumes of convex sets', *Mathematika*, **7**, 125-138 (1960)
93. Fenchel, W. and Jessen, B., 'Mengenfunktionen und konvexe Körper', *Danske Videnskabernes Selskab. Matematisk-Fysiske Meddelelser*, **16**, 1-31 (1938)
94. Schneider, R., 'Kinematische Berührmasse für konvexe Körper und Integralrelationen für Oberflächenmasse', *Mathematische Annalen*, **218**, 253-267 (1975)

95. Federer, H., 'Curvature measures', *American Mathematical Society. Transactions,* **93,** 418-491 (1959)

96. Firey, W.J. and Grünbaum, B., 'Addition and decomposition of convex polytopes', *Israel Journal of Mathematics,* **2,** 91-100 (1964)

97. Firey, W.J., 'Blaschke sums of convex bodies and mixed bodies' in *Proceedings of the colloquium on convexity, Copenhagen, 1965* (ed. W. Fenchel), Københavns Universitets Matematiske Institut, Copenhagen (1967)

98. Efimov, N.V., *Flächenverbiegung im Grossen,* Akademie Verlag, Berlin (1957); translated from Russian

99. Pogorelov, A.V., *Extrinsic geometry of convex surfaces,* Translations of Mathematical Monographs 35, American Mathematical Society, Providence, R.I. (1973); translated from Russian

100. Pogorelov, A.V., *Die eindeutige Bestimmung allgemeiner konvexer Flächen,* Akademie Verlag, Berlin (1956); translated from Russian

101. Pogorelov, A.V., *Die Verbiegung konvexer Flächen,* Akademie Verlag, Berlin (1957); translated from Russian

102. Matheron, G., *Integral geometry and geometric probability,* Wiley, New York (1975)

103. Larman, D.G., and Rogers, C.A., 'The existence of a centrally symmetric convex body with central sections that are unexpectedly small', *Mathematika,* **22,** 164-175 (1975)

104. Schneider, R., 'Zu einem Problem von Shephard über die Projektionen konvexer Körper', *Mathematische Zeitschrift,* **101,** 71-82 (1967)

105. Kendall, M.G. and Moran, P.A.P., *Geometric probability,* Griffin, London (1963)

106. John, F., 'Extremum problems with inequalities as subsidiary conditions' in *Studies and essays presented to R. Courant* (ed. K.O. Friedrichs, O.E. Neugebauer and J.J. Stoker), Wiley, New York (1948)

107. Danzer, L., Laugwitz, D. and Lenz, H., 'Über das Löwnersche Ellipsoid und sein Analogen den einem Eikörper einbeschriebenen Ellipsoiden', *Archiv der Mathematik,* **8,** 214-219 (1957)

108. Grünbaum, B., 'Borsuk's problem and related questions', *American Mathematical Society. Proceedings of Symposia in Pure Mathematics,* **7,** 271-284 (1963)

109. Knast, R., 'An approximate theorem for Borsuk's conjecture', *Cambridge Philosophical Society. Proceedings,* **75,** 75-76 (1974)

110. Bang, T., 'A solution of the "Plank Problem"', *American Mathematical Society. Proceedings,* **2,** 990-993 (1951)

13

Topology

*J.F. Adams and A.R. Pears**

13.1 INTRODUCTION

The object of this section is to give guidance on, and help with, the use of the literature in two fields, which can broadly be described as (1) classical, homotopy-invariant algebraic topology, and (2) its application to the study of manifolds. The former is a subject which enjoyed a period of rapid growth starting in the early 1950s; it now presents a well-developed body of theory, and seems to be moving more slowly. By contrast there is now (1975) more research activity in the application of algebraic topology to problems arising elsewhere, and here one must include almost all problems about manifolds. Of course, on the one hand, problems about manifolds do not exhaust the applications of algebraic topology, and, on the other hand, their solution may require other techniques in addition to those of algebraic topology. Still, it would seem hard to do serious work on manifolds without a knowledge of the classical methods of algebraic topology.

The primary sources in these areas are of course papers in the learned journals; and these can be located in the usual way via *Mathematical Reviews* and similar bibliographical works (cf. Section 3.4). Of particular use here are the *Reviews of papers in algebraic and differential topology, topological groups and homological*

----

* J.F. Adams contributed the sections on algebraic topology and manifolds; A.R. Pears, the section on general topology [editor].

algebra, edited by N.E. Steenrod, and published in two volumes by the American Mathematical Society, Providence, R.I. (1968). Unfortunately it only goes up to 1967, and this in the sense that if a paper appeared in 1967 and its review appeared in 1968, then it will not be included. However, there are now (1977) plans to bring it up to date under the editorship of A. Clark.

The expert will require no further guidance; and the layman should be directed to the articles in the *Encyclopaedia Britannica.* Editions before 1974 carried an article on algebraic topology by Prof. W.S. Massey which is basically excellent, but now shows its age a little. A new article was commissioned for the 1974 edition; this does cover more up-to-date material, but the coverage attempted is perhaps a little ambitious for complete understanding by a lay reader. With these two classes of reader dismissed, I may address the main body of readers, whom I suppose to have at least a first degree in mathematics, but not to be expert in this particular area.

13.2   ALGEBRAIC TOPOLOGY

So far as algebraic topology goes an attempt has been made to give the necessary guidance and help in J.F. Adams, *Algebraic topology; a student's guide,* Cambridge University Press (1972; London Mathematical Society Lecture Note Series 4). In that book I begin with a 31-page survey of the material facing the student of algebraic topology, and commenting on the sources from which it can most conveniently be studied; the book continues as a reprint collection. The introductory survey, of course, is more full than I can manage in the present space, so I refer the reader to it. It seems to wear reasonably well. However, because of publication difficulties, the references were not up to date when the book appeared; and, in particular, the list of books on pp. 1-4 contains nothing later than 1967. Today it would be advisable to add a number of books to this list. The following may all be considered as 'first textbooks', but have different slants:

E. Artin and H. Braun, *Introduction to algebraic topology,* Charles E. Merrill Publishing Co, Columbus, Ohio (1969)
C. Godbillon, *Éléments de topologie algébrique,* Hermann, Paris (1971)
A. Gramain, *Topologie des surfaces,* Presses Universitaires de France, Paris (1971)
C.R.F. Maunder, *Algebraic topology,* Van Nostrand, Reinhold, London (1970)
J. Mayer, *Algebraic topology,* Prentice-Hall, Englewood Cliffs, N.J. (1972)

C.T.C. Wall, *A geometric introduction to topology*, Addison-Wesley, Reading, Mass. (1972)

A.H. Wallace, *Algebraic topology: homology and cohomology*, Benjamin, New York (1970)

Wall's book can be recommended for its individual approach. Otherwise, the book by Maunder is the most comprehensive, and should probably be recommended. In addition, the old book on *Algebraic topology* by W. Franz has been translated (Frederick Ungar, New York, 1968); but there seems no particular reason to recommend it in view of later texts.

Of the books recently published, the following two seem the most useful as 'second textbooks': A. Dold, *Lectures on algebraic topology*, Grundlehren der Mathematischen Wissenschaften 200, Springer-Verlag, Berlin (1972); R.M. Switzer, *Algebraic topology – homotopy and homology*, Grundlehren der Mathematischen Wissenschaften 212, Springer-Verlag, Berlin (1975).

Dold has an excellent style, but the subject matter of his book is a little restricted, centering around classical homology. Switzer's book covers a wider range of material than Dold's, and indeed is very substantial; apart from its price, which is excessive, it is probably the most recommendable choice as a second textbook.

The following books combine expository intent and specialist content in various proportions:

J.F. Adams, *Stable homotopy and generalised cohomology*, University of Chicago Press, Chicago (1974)

R. F. Brown, *The Lefschetz fixed point theorem*, Scott, Foresman and Co, Glenview, Ill. (1971)

G.E. Cook and R.L. Finney, *Homology of cell complexes*, Princeton University Press, Princeton (1967)

E. Dyer, *Cohomology theories*, W.A. Benjamin, New York (1969)

F. Harary, *Graph theory*, Addison-Wesley, Reading, Mass. (1969) (Harary's is the standard book on this topic.)

P.J. Hilton (ed.), *Studies in modern topology*, MAA Studies in Mathematics 5, Mathematical Association of American, Buffalo, N.Y. (1968) (This contains interesting expository surveys.)

P.J. Hilton, G. Mislin and J. Roitberg, *Localisation of nilpotent groups and spaces*, Mathematics Studies 15, North-Holland, Amsterdam (1975) (This is essentially a specialist monograph; but localisation is important.)

S.T. Hu, *Homology theory*, Holden-Day, San Francisco (1966)

S.T. Hu, *Cohomology theory*, Markham, Chicago (1968) (In general, books by Hu have no advantage over the sources from which they derive.)

S.Y. Husseini, *The topology of the classical groups and related topics,* Gordon and Breach, New York (1969)

K. Lamotke, *Semisimpliziale algebraische Topologie,* Grundlehren der Mathematischen Wissenschaften 147, Springer-Verlag, Berlin (1968)

A.T. Lundell and S. Weingram, *The topology of CW-complexes,* Van Nostrand Reinhold, London (1969)

J.W. Milnor and J.D. Stasheff, *Characteristic classes,* Annals of Mathematics Studies 76, Princeton University Press, Princeton (1974) (Milnor's notes on the subject date from 1957; their appearance in print has been long awaited, and is most welcome.)

R.E. Mosher and M.C. Tangora, *Cohomology operations and applications in homotopy theory,* Harper and Row, New York (1968) (For getting a certain sort of know-how this is definitely readable. The trouble is that it may encourage students to do unintelligent calculations in an area where unintelligent calculations are the last thing we need.)

J. Stallings, *Group theory and three-dimensional manifolds,* Yale Mathematical Monographs 4, Yale University Press, New Haven, Conn. (1971)

R. Stong, *Notes on cobordism theory,* Princeton University Press, Princeton (1968) (This is the most useful standard reference work on cobordism.)

13.3 MANIFOLDS

I now turn to the subject of manifolds. Here one might remark that many subjects seem to pass rather rapidly from a state in which there are too few books to one in which there are too many; and large areas of manifold theory (if not the whole of it) are still in the former state. So, in a sense, there are fewer 'resources' to be surveyed than in the last section.

It is convenient to subdivide our topic, according to the nature of the assumptions made on the manifolds, into the study of topological manifolds, piecewise-linear manifolds and differentiable (or smooth) manifolds. Of course, there are large bodies of work which seek to relate manifolds, as defined under one set of rules, to manifolds defined under another set of rules; for example, one may take a given piecewise-linear manifold and enquire how many smooth structures (if any) may be put on it. Indeed, it is only recently that it has been *proved* that these differing assumptions lead to different theories, in the sense that there are manifolds which satisfy one assumption but not another; it seems that some of the pioneers hoped for the contrary state of affairs, so that the three theories would have been equivalent; this would no doubt have been simpler, but it turns out that the state of affairs is not so simple.

It is perhaps natural to begin with differentiable (smooth) manifolds, for those which arise in the rest of mathematics are most often of this type. The assumptions made on the manifolds are such that one can do differential geometry. Therefore the basic justification for the geometric constructions which are made comes from analysis; but one can perform 'obvious' geometric constructions quite freely, provided that one avoids non-differentiable means which might introduce kinks and corners. The theory of fibre bundles provides the most important single tool, and often allows one to reduce geometrical questions to problems which can be solved by algebraic topology.

The following are among the books which might serve as a starting point:

S. Lang, *Differential manifolds*, Addison-Wesley, Reading, Mass. (1972) (This is a revised and greatly expanded version of the author's earlier book, *Introduction to differential manifolds*, John Wiley, New York, 1962. It presents the way an expert would like to see the foundations set up – but parts of it may be a bit abstract for the beginner.)

J.W. Milnor, *Topology from the differentiable viewpoint*, University Press of Virginia, Charlottesville (1965) (Recommended.)

J.R. Munkres, *Elementary differential topology*, Annals of Mathematics Studies 54, Princeton University Press, Princeton (1963)

One more advanced book has been found useful by many students:

J.W. Milnor, *Lectures on the h-cobordism theorem*, Princeton University Press, Princeton, N.J. (1965)

Then we have the following survey articles:

J.W. Milnor, 'Differential topology' (pp. 165-183 of T.L. Saaty (ed.), *Lectures on modern mathematics*, Vol. 2, Wiley, New York, 1964)

S. Smale, 'A survey of some recent developments in differential topology', *American Mathematical Society. Bulletin*, **69**, 131-145 (1963)

C.T.C. Wall, 'Topology of smooth manifolds', *London Mathematical Society. Journal*, **40**, 1-20 (1965)

These are strongly recommended; and of course they have references. However, I will cite two papers which I feel mark the birth of modern differential topology:

J.W. Milnor, 'On manifolds homeomorphic to the 7-sphere', *Annals of Mathematics*, **64**, 399-405 (1956)

S. Smale, 'Generalised Poincaré conjecture in dimensions greater than four', *Annals of Mathematics*, **74**, 391-406 (1961)

The paper by Milnor is the one which first showed that the category of smooth manifolds is different from the other two categories. The paper by Smale is important not only for its result, but also for its method, which I may summarise as follows. In algebraic topology one usually begins by studying finite simplicial complexes; these are spaces which can be subdivided into points, line-segments, triangles and their higher-dimensional analogues, which are called simplexes. However, it turns out to be more efficient to study CW-complexes, which are made up out of 'cells' instead of simplexes; one needs only a small number of cells compared with a large number of simplexes, and it turns out that (under suitable assumptions) a decomposition into cells can be made to follow the topological invariants of the space very closely. The analogue for a manifold of a cell-decomposition is a decomposition into 'handles'. It is important to study such decompositions for piecewise-linear and topological manifolds also, but they were originally introduced for smooth manifolds; the idea probably arose from modern interpretations, owing much to R. Bott, of the work of M. Morse. A decomposition into handles can be rearranged and manipulated; and as for cells, it turns out that (under suitable assumptions) a decomposition into handles can be made to follow the topological invariants of a manifold very closely. This is Smale's method.

Somewhat related to handle-decompositions is the method of 'surgery'. In this we take a manifold and cut out entirely a suitable part of it (which for present purposes we may think of as being a little more than one 'handle'). We then glue in a new part, so obtaining a new manifold, different from but related to the one we started with. R. Thom's concept of 'cobordism' may be interpreted in this way; and the method is very useful. The original reference is as follows: J.W. Milnor, 'A procedure for killing homotopy groups of differentiable manifolds', *American Mathematical Society. Proceedings of Symposia in Pure Mathematics,* **3**, 39-55 (1961).

There are two books: W. Browder, *Surgery on simply-connected manifolds,* Ergebnisse der Mathematik und ihrer Grenzgebiete 65, Springer-Verlag, Berlin (1972); and C.T.C. Wall, *Surgery on compact manifolds,* London Mathematical Society Monographs 1, Academic Press, London (1970). The student should start on Browler's: Wall's is harder.

I turn now to the topology of piecewise-linear manifolds. These are ones which can be made up from points, line-segments, triangles, and more generally simplexes, assembled in a prescribed way. The basic justification for the geometric constructions which are made now comes from elementary linear algebra. Kinks and corners, which were taboo in the smooth theory, are now the order of the day. One can perform 'obvious' geometric constructions quite freely. Contrasting this theory with that of topological manifolds (where one has to construct some fairly pathological homeomorphisms), one is tempted to say that in the piecewise-linear case the 'obvious' constructions are the only ones; but the reader who pauses to consider (for example) knots in 4, 5 and 6 dimensions will see that genuine geometric insight is required, and in that sense the constructions may be far from obvious. Once one gets past the elementary results which can be proved by 'general-position' arguments, proofs tend to proceed by the inductive repetition of elementary steps or moves; this method is very appropriate because of the finitistic nature of the material. This finitistic or combinatorial character of the subject probably helped to attract and encourage the early workers, and the subject has a history going back to the 1930s. The reader may like the following survey article by one who helped to shape that history: M.H.A. Newman, 'Geometrical topology', *International Congress of Mathematicians. Proceedings*, **9**, 139-146 (1962).

At the beginning of the 'modern' period the following notes had a considerable influence: E.C. Zeeman, *Seminar on combinatorial topology*, mimeographed notes, Institut des Hautes Études Scientifiques, Paris (1963; Chapter 7 revised 1965, Chapter 8 revised 1966). Unfortunately these are not easily available. However, the material is mostly available in book form:

L.C. Glaeser, *Geometrical combinatorial topology*, Vol. 1, Mathematical Studies 27, Van Nostrand Reinhold, London (1970)
J.F.P. Hudson, *Piecewise-linear topology*, Benjamin, New York (1969)
C.P. Rourke and B.J. Sanderson, *Introduction to piecewise-linear topology*, Ergebnisse der Mathematik und ihrer Grenzgebiete 69, Springer-Verlag, Berlin (1972)
J.R. Stallings, *Lectures on polyhedral topology*, Tata Institute of Fundamental Research, Bombay (1968)

The first is shorter than the other three, and from a slightly different tradition; the other three seem more influenced by Zeeman. The fourth is not quite so easily available as the others; it has some nice material near the end, but there are too many pages to read before you get to it. Presumably one recommends the second and third.

One research monograph should be mentioned here: M.W. Hirsch and B. Mazur, *Smoothings of piecewise-linear manifolds*, Annals of Mathematics Studies, 80, Princeton University Press, Princeton, N.J. (1974).

In certain situations, the question of what can be done by a finite number of elementary steps or moves can be reduced to pure algebra. This brings us to the theory of 'torsion' or 'simple-homotopy type'.

Here the word 'simple' is not to be taken in its everyday sense; the theory is no more simple than the theory of 'simple groups'. However, there is a theory, and it is useful both in piecewise-linear and in smooth topology. I would direct the reader firmly to Milnor's excellent exposition: J.W. Milnor, 'Whitehead torsion', *American Mathematical Society. Bulletin,* 72, 358-426 (1966).

There are also two accounts available in book form: M.M. Cohen, *A course in simple-homotopy theory,* Graduate Texts in Mathematics series 10, Springer-Verlag, Berlin (1973); and G. de Rham, S. Maumary and M.A. Kervaire, *Torsion et type simple d'homotopie,* Lecture Notes in Mathematics 48, Springer-Verlag, Berlin (1967).

In all three categories, studies of the 'position' of a subset in a manifold are fundamental. The problem reaches perhaps its most characteristic form in the study of knots. There is one canonical book on 'ordinary' knots: R.H. Crowell and R.H. Fox, *Introduction to knot theory,* Ginn, Boston (1963).

However some feel that one can get the same ideas by reading fewer words in the following survey article: R.H. Fox, 'A quick trip through knot theory' (pp. 120-167 of M.K. Fort (ed.), *Topology of 3-manifolds and related topics,* Prentice-Hall, Englewood Cliffs, N.J., 1962). This is perhaps the recommended source.

Of course, from the point of view of the rest of mathematics, knots in higher-dimensional space deserve just as much attention as knots in 3-space. On this topic I am reduced to citing, with some misgivings, a selection of the original sources:

A. Haefliger, 'Knotted $(4k - 1)$-spheres in $6k$-space', *Annals of Mathematics,* 75, 452-466 (1962)

E.C. Zeeman, 'Unknotting combinatorial balls', *Annals of Mathematics,* 78, 501-526 (1963)

M.A. Kervaire, 'Les noeuds de dimension supérieures', *Société Mathématique de France. Bulletin,* 93, 225-271 (1965)

J. Levine, 'Unknotting spheres in codimension two', *Topology,* 4, 9-16 (1965)

J. Levine, 'A classification of differentiable knots', *Annals of Mathematics,* 82, 15-50 (1965)

Experts in the subject tell their students to read Levine.

Finally we come to the subject of topological manifolds. Here the basic justification for the geometric constructions which are made must come from general topology — that is, 'analytic' or 'point-set' topology; and indeed the subject draws much of its flavour and much of its impetus from that source. More precisely, in order to make any use of the difference between the topological category and the smooth or piecewise-linear category, one must construct maps (and, more especially, homeomorphisms) which are not locally 'good' at all, but to smooth or piecewise-linear eyes present the most extreme singularities. Inescapably the methods for handling such things have much in common with those for constructing and handling 'pathological examples' in general topology; but these days they also require extensive awareness of algebraic topology and of developments elsewhere in the theory of manifolds (surgery obstructions, handlebodies).

We may regard the distinctive flavour of the subject as beginning to emerge with the work of Moise and, particularly, Bing; but the many works of these authors are not all easy to read. Another choice for a paper making the birth of the 'modern' period might be the following: M. Brown, 'A proof of the generalised Schoenflies theorem', *American Mathematical Society. Bulletin,* **66**, 74-76 (1960).

There has been much activity in this field recently; in particular, the student should be aware that Kirby and Siebenmann have made important progress, and have (for example) proved the Annulus Conjecture. There is exactly one book: T.B. Rushing, *Topological embeddings,* Academic Press, New York (1973). This therefore becomes the recommended source; it has useful references.

Finally, it remains to comment on volumes which contain the proceedings of various specialist conferences. Conferences have the drawback that they tend to be ephemeral; who would want to attend last year's conference, if it were available perfectly recorded on videotape? They have the virtue that they give a vivid impression of mathematics as a living enterprise. The proceedings of conferences have both the drawback and the virtue in a diluted form. It often happens that the best contributions to conferences are also published properly in the usual journals; and to this extent conference proceedings present the disadvantage that they duplicate other journals, but appear in an unsystematic way, at irregular intervals, and under a new editor and title each time. One might be excused for consulting them only when a bibliographical search reveals that some relevant paper appeared in one. However, there are a few conferences which, by a fortunate choice of subject at a fortunate time, publish proceedings with a higher ratio than usual of contributions which one wishes to consult; and these may deserve a place on one's bookshelf. Perhaps the best example in this area is:

M.K. Fort (ed.), *Topology of 3-manifolds and related topics,* Prentice-Hall, Englewood Cliffs, N.J. (1962)

For the rest, the most recent reference may be the most useful, as later references allow you to find earlier ones but not vice versa. So as a sample of conference proceedings in general I cite a recent one: L.F. McAuley (ed.), *Algebraic and geometrical methods in topology,* Lecture Notes in Mathematics 428, Springer-Verlag, Berlin (1974).

13.4   GENERAL TOPOLOGY

Topology is a rather diverse subject with many origins. General topology consists in the main of the study of abstract spaces and mappings between them but also includes many other topics which do not belong to the areas of algebraic topology, differential topology and global analysis. Soon after Cantor has initiated the theory of sets, Fréchet began, in 1906, the study of abstract spaces. The concept of topological space soon evolved. Point-set topology is the study of general topological spaces and continuous mappings. Some knowledge of point-set topology is essential for all work in mathematics. The basic notions and general constructions should be known. The most interesting classes of spaces for non-specialists are the compact spaces and the metrisable spaces. The class of paracompact spaces contains both of these classes and has turned out to be the 'right' class of spaces for many purposes. The non-specialist should perhaps also know the Nagata–Smirnov theorem which characterises the topological spaces which are metrisable. Textbooks giving an exposition of this fundamental material are: J. Dugundji, *Topology*, Allyn and Bacon, Boston (1966); R. Engelking, *Outline of general topology*, North-Holland, Amsterdam (1968) and J.L. Kelley, *General topology*, Van Nostrand, Princeton, N.J. (1955).

Very many generalisations of the classes of paracompact and metrisable spaces have been introduced. It is not possible to say which of these classes will be found ultimately to be significant. A unified approach to the problems of classification of spaces and mappings was developed by A.V. Arkhangel'skii, 'Mappings and spaces', *Russian Mathematical Surveys*, 21, 115-162 (1966). A detailed account of the ring of continuous real-valued functions on a topological space is given by L. Gillman and M. Jerison, *Rings of continuous functions*, Van Nostrand, Princeton, N.J. (1960). The algebraic structure of this ring gives information about topological properties of the space. A closely connected topic is the Stone–Čech compactification, and this is also investigated in detail by Gillman and Jerison.

Peano's example of a space-filling curve forced questioning of the meaning of dimension. The problem of distinguishing topologically between different Euclidean spaces was the starting point of dimension theory. The classical dimension theory of separable metric spaces is elegantly exposed by W. Hurewicz and H. Wallman, *Dimension theory*, Princeton University Press, Princeton, N.J. (1941). There is no satisfactory single extension of the theory of dimension to general topological spaces. For an account of the various dimension theories for non-metrisable spaces, the relations between them, and examples showing the pathological aspects of the theories see: K. Nagami,

Dimension theory, Academic Press, New York (1970) and A.R. Pears, *Dimension theory of general spaces,* Cambridge University Press, Cambridge (1975).

Although topological spaces and continuous mappings are the main concern of point-set topology, other types of 'continuity structure' are studied. The theory of uniform spaces is analogous to the theory of metric spaces but is of wider applicability. This is the setting in which the concept of uniform continuity can be most naturally investigated. There are two approaches to uniformity: by means of uniform covering, the theory being developed from this point of view in J.R. Isbell's book, *Uniform spaces,* Mathematical Surveys 12, American Mathematical Society, Providence, R.I. (1964); and by means of certain relations, called entourages, which are employed by N. Bourbaki in Chapter 2, 'Structures uniformes', of his *Topologie générale,* Hermann, Paris (1961) (translated as *General topology,* 2 vols, Addison-Wesley, Reading, Mass., 1966). In a uniform space there is a notion of 'nearness' of sets and this can be abstracted to provide the definition of a proximity space. An introduction to the theory of proximity spaces and their generalisations is given by S.A. Naimpally and B.D. Warrack, *Proximity spaces,* Cambridge Tracts in Mathematics and Mathematical Physics 59, Cambridge University Press, Cambridge (1973). E. Čech, *Topological spaces,* Publishing House of Czechoslovak Academy of Sciences, Prague; Interscience, London (1966) is an interesting presentation of topologies, uniformities and proximities which is completely self-contained, all necessary mathematical concepts, beginning with class and set, being introduced in the work. A.W. Hager's paper entitled 'Some nearly fine uniform spaces', *London Mathematical Society. Proceedings,* **28,** 517-546 (1974) contains a bibliography of recent work on uniform spaces.

In 1963 P.J. Cohen proved that it is consistent with the usual axioms for set theory that the continuum hypothesis be false. The technique of proof, called forcing, which Cohen introduced, has recently been applied to many questions in general topology of a set-theoretic nature. Many such problems have a translation into questions of cardinal arithmetic; the text by I. Juhász, 'Cardinal functions in topology', *Mathematical Centre Tracts,* **34** (1971), gives much information on this subject. There is at present much interest in the answers to topological questions in special models for set theory — in particular, in the contrasting models (a) satisfying Martin's axiom together with the negation of the continuum hypothesis and (b) Gödel's constructable universe (in which the generalised continuum hypothesis holds). For a survey of the whole area affected by set-theoretic influence and of the methods used see: M.E. Rudin, *Lectures on set theoretic topology,* Regional Conference Series in Mathematics 23, American Mathematical Society, Providence, R.I. (1975).

Finally a selection should be made from the numerous topics in addition to point-set topology which belong to the field of general topology. Many topological results of a general nature can be described and analysed in categorical terms. The notes of H. Herrlich, *Topologische Reflexionen und Coreflexionen*, Lecture Notes in Mathematics 78, Springer-Verlag, Berlin (1968), provide an introduction to categorical topology. The shape of a topological space is a modification of its homotopy type. The articles by S. Mardešić, 'A survey of the shape theory of compacta', *Prague Topological Symposium. Proceedings*, **3**, 291-300 (1971) and 'Shapes for topological spaces', *General Topology and its Applications*, **3**, 265-282 (1973), give the basic definitions of this rapidly developing field. There does not seem to be an expository account of infinite dimensional topology, but the report *Symposium on Infinite Dimensional Topology*, edited by R.D. Anderson (Annals of Mathematics Studies 69, Princeton University Press, Princeton, N.J., 1974), should indicate the nature of this field.

14

Mathematical Programming

J.M. Brown

14.1 INTRODUCTION

Mathematical programming is constrained optimisation. The abstract problem is to find the greatest or least value of a function of many variables with the variables constrained to some subdomain of R_N. Usually, they are non-negative and satisfy other inequalities, and may be further limited to say integer values.

However, this is very much an applied branch of mathematics, recognised as such by the 1975 Nobel awards in economics to Kantorovitch[1] and Koopmans[2] for their early transportation studies. Existence, uniqueness and characterisation of optima are important, but far more emphasis is given to their calculation. The subject developed in parallel with digital computers, and for a technique to become established, reliable and efficient computer program implementations are necessary. This emphasis is important and different from most other chapters of this volume.

The subject is well covered by textbooks, but journal articles and conference proceedings are often very specialised. These are scattered under many headings: operational research; management; economics; military strategy; computing; combinatorics; optimisation; chemical, structural and electrical engineering! This survey is far from comprehensive, but attempts to locate comprehensible treatments, key works in the field and unusually illuminating articles on difficult aspects.

References are cited by author(s), and listed in full at the chapter end. Generally available, English-language editions are given wherever possible. Linear programming is covered first (cf. Section 14.5), then generalisations to non-linear versions with continuous variables. Unconstrained optimisation is briefly mentioned as it has undergone similar development. Problems with discrete variables are then included, although this field is very new and involves concepts closer to computing than to traditional mathematics.

14.2 REVIEWS AND GENERAL COVERAGE BOOKS

Few books cover the whole of this field, and many are given in the context of applications. However, McMillan[3] stands out at the introductory level for broad coverage and few prerequisites. The survey by Wolfe (Chapter VI in Abadie[4]) is very readable, and so are his comments in the introduction to Fletcher[5]. Dorn[6] has dated somewhat. Wilde and Beightler[7] attempt a unified presentation, although it is rather large. Similarly, Hadley[8,9] covers most of the material thoroughly, though from an introductory viewpoint.

More advanced wide-ranging presentations are Zoutendijk[10], Zangwill[11] and Mangasarian[12]. Zoutendijk[10] and Beale (Chapter VII in Abadie[4]) survey computational aspects, and Chapter 3 of Saaty and Bram[13] has this emphasis, while Goldstein[14] gives an interesting advanced presentation. Geffrion[15,16] is extensive and gives a useful bibliography.

From the applications in operational research viewpoint, Ackoff and Sasieni[17] is a popular student text, Saaty[18] and Vajda[19] extensive but dated and Mitchel[20] good and unusual in being written by a practitioner (National Coal Board). Nicholson[21] is also industrially oriented and is good on the new combinatorial aspects.

14.3 BIBLIOGRAPHY, ABSTRACTS AND RETRIEVAL SYSTEMS

For the narrow field of mathematical programming, the main bibliographies appear to be Leon in pp. 599-649 of Reference 22, Geffrion[16] and Riley and Gass[23] in English, and Kunzi and Oettli[24] in German. However, operational research is better covered by the comprehensive but poorly indexed Batchelor[25] and useful Buckland[26].

Many abstracting journals include sections on this field, and some major ones are listed below.

International Abstracts in Operational Research
Operations Research/Management Science Abstracts Service
Computer and Control Abstracts (Institution of Electrical Engineers, Science Abstract C)
Computer Abstracts
Computing Reviews
Mathematical Reviews
Chemical Abstracts

Computer-based information retrieval systems are useful for clear enquiries by author or subject keywords (beware of ambiguous words such as 'programming'). Of these, the INSPEC (*Science Abstracts*) system seems most appropriate, though the *Engineering Index* and *Chemical Abstracts* will again hold relevant material. Although too expensive to allow misuse, such systems are especially valuable for recent literature surveys and tracing vaguely remembered articles. They are not difficult to use.

14.4 PERIODICALS

Papers on mathematical programming have appeared in so many journals that only the major current ones are given here. Only one seems to be wholly devoted to the topic: *Mathematical Programming* (North Holland, 1971– ; in English). Others are listed below with occasional comments under their usual subject headings.

14.4.1 Mathematics

Institute of Mathematics and its Applications. Journal (1965–).
  Now taking the numerical computation papers previously published by the *Computer Journal* (cf. Subsection 14.4.2).
Journal of Optimization Theory and Applications (Plenum, 1967–)
Numerische Mathematik (Springer, 1959–). Much of it in English or with English summaries.
SIAM Journal on Control (1963–)
SIAM Review (1959–)
Society for Industrial and Applied Mathematics. Journal, **1–13** (1953-1965); continued as *SIAM Journal on Applied Mathematics*, **14–** (1966–).

Society for Industrial and Applied Mathematics, Journal. Series B: numerical analysis 1–2 (1964-1965); continued as *SIAM Journal on Numerical Analysis,* 3– (1966–   ).

14.4.2 Computing

Association for Computing Machinery. Communications (1958–   )
Association for Computing Machinery. Journal (1954–   ). The former is very popular reading, and the latter more detailed.
Computer Journal (British Computer Society, 1958–   ). Now emphasising non-numerical computation.
Computing (Springer, 1966–   ). Partly in English.

14.4.3 Operational research, economics, management, etc.

Econometrica (Econometrica Society, 1933–   )
Journal of Aerospace Sciences (American Institute of Aeronautics and Astronautics, 1960-1962); continued as *A.I.A.A. Journal* (1963–   )
Management Science (Institute of Management Sciences, 1954–   )
Naval Research Logistics Quarterly (United States: Navy Department. Office of Naval Research, 1954–   )
Operational Research Quarterly (Operational Research Society, 1950–   ).
Operations Research Society of America. Journal (1952-1955); continued as *Operations Research* (1956–   ).

Journals of some large organisations such as IBM, Bell Telephones, Rand Corporation, National Coal Board and British Steel Corporation.

14.5 LINEAR PROGRAMMING

This is the optimisation of a linear function subject to linear equality and inequality constraints. The theory is complete, well documented, and basic to almost all non-linear programming. Solution is via a version of Dantzig's 'Simplex algorithm', and computer programs implementing this are available at all respectable computer centres.

At an elementary level, most modern applied algebra texts include the topic. The text by Kemeny *et al.*[27] is typical for business-oriented mathematics or statistics. Science students find Chapter 6 in Noble[28] particularly readable. Trustrum[29] is concise and cheap. More comprehensive treatments are given by Hadley[8], Gass[30] and Charnes, Cooper and Henderson[31], although there are numerous other books at this level. Dantzig[32] is thorough, and Gale[33] unusually elegant and readable despite the economics context. Kuhn and Tucker[34] survey the pure mathematical aspects with convex polytopes and duality in the forefront. Duality — that to each minimisation problem there is a corresponding dual maximisation problem — is of more than theoretical interest.

The 'Assignment and Transportation Problems' are linear programs of special form allowing simplification. They are related and explained in context in most books (e.g. Gale[33]). Solution may be via networks, which are covered later (cf. Section 14.11), or Kuhn's[35] Hungarian method, which is more simply explained by Flood[36]. Here the models have links with abstract transversal and matroid theory (see Gale[37]).

The subject arose from Leontief's 1941 linear economic model[38] for which von Neumann and Morgenstern[39] developed competitive models in their 'Theory of Games'. This application of linear programming is given in most of the books, as it provides the optimal strategies.

14.6 GENERAL NON-LINEAR PROGRAMMING

Here the function or constraints, or both, are non-linear. The optimum may lie on some constraints with the inequalities attaining equality, or may be away from all the inequality constraints. Kuhn and Tucker extended the theory of Lagrange multipliers to cover these situations. This is simply explained in McMillan[3], and more rigorously treated in Hadley[9].

Uniqueness of the optimum is difficult to treat in general, and only the case of convex functions and domains is thoroughly understood. Much of the theory of linear programming readily generalises to this case of 'convex programming'. Most non-linear programming texts include some version of this, and a particularly elegant treatment is given in the first half of Berge and Ghouila-Houri[40].

The topic is surveyed by Wolfe (Chapter VI of Abadie[4]), and further contributions are given in Rosen *et al.*[41] and Abadie[42]. Saaty and Bram[13] treat this together with unconstrained optimisation, as was the unifying aim of the conference edited by Fletcher[5].

Quadratic programming is the special case of minimising a quadratic positive definite (and so convex) function subject to linear constraints. The most effective algorithm is that of Beale (see Chapter VII of Abadie[4]), and a survey of algorithms is given by Boot[43].

A multitude of algorithms have been proposed and their relative merits seem undecided. Limited comparisons are offered by Colville[44] and Leon in Reference 22 (pp. 28-46). Rosen[45, 46] deserves special mention, as do Goldfarb and Lapidus[47], who solved particularly difficult real problems. However, with the development of new techniques such as that described next, the whole efficiency emphasis may change.

14.7 GEOMETRIC PROGRAMMING

The name of this technique derives from the geometric mean-arithmetic mean inequality (Cauchy's) for positive terms. At equality, the minimised arithmetic mean equals the maximised geometric mean, so they are dual quantities in the mathematical programming sense. From such beginnings noticed by Zener, Duffin and Peterson generalised this into a programming technique described in Duffin, Zener and Peterson[48]. Further limitations were overcome by Wilde and Passy (see Wilde and Beightler[7]). This is well reviewed in Peterson[49], where the extraordinary versatility of the technique is emphasised.

To explain the wide range of problems covered by this technique, 'posynomials' must first be defined. These are weighted sums of terms consisting of products of the variables to given powers, not necessarily integral. Geometric programming minimises such posynomials subject to inequality constraints of posynomial form.

There are a few limitations: convexity needed to assure a global minimum, and 'degrees of difficulty' explained in the references. However, the method seems ideal for applications, especially as industrial plant costs are usually given in posynomial form (the terms found from logarithmic graphs). The full implications of this technique have not yet been realised.

14.8 UNCONSTRAINED OPTIMISATION

This section is included as the pure mathematician may not think there is a problem at all! Mathematical programming (constrained) problems can also be converted to this form.

When a function is unconstrained, its minimum is characterised by a zero gradient and positive definite Hessian matrix. Convexity assures a global minimum. All seems easy, except for the complicated functions arising in present-day applications. With non-quadratic functions, the gradient equations are non-linear and their solution is only just systematised (Ortega and Rheinbolt[50], who also cover optimisation and give many references).

In fact, direct minimisation is usually preferred, and the considerable developments beyond steepest descent are ably reviewed by Powell[51]. The book by Jacobi, Kowalik and Pizzo[52] summarises these, and gives a bibliography including computer programs. Box, Davies and Swann[53] cover the material from an industrial viewpoint. Simple inequality constraints can often be eliminated by a change of variable, e.g. with x in $(-1, 1)$ let $x = \tanh z$, then z (real) is unlimited. These are useful in theory as well as applications, and are mentioned by Box[54]. Another approach uses a sequence of Lagrange multiplier values handling equality and inequality constraints. The resulting 'Sequential Unconstrained Minimisation Technique' is described by Fiacco and McCormick[55], and has solved some very complicated engineering problems.

14.9 DYNAMIC PROGRAMMING

This is an approach to optimisation rather than a particular technique, and is due in a unified form to Bellman[56]. Basically, it decomposes an optimisation problem into a sequence of subproblems satisfying an 'optimality principle'. This is related to the calculus of variations (Dreyfus[57] and Bellman[56]), and to the Pontryagin optimality principle more suited to differential equation formulation (Pontryagin *et al.*[58]).

In mathematical programming, dynamic programming has the advantage of applying to *discrete* combinatorial problems which are common in operational research, and are emphasised in the remainder of this chapter.

For example, the obvious algorithm for expressing an amount of money with fewest coins may be derived in this way. The approach seems best explained by examples, and the books by Hadley[9], Nemhauser[59], Bellman and Dreyfus[60] and Kaufmann and Cruon[61] give many. Multistage systems represented by boxes joined with arrows are particularly amenable to treatment, as is emphasised in Chapter 8 of Wilde and Beightler[7].

For other problems, the approach is difficult to formulate, and may lead to algorithms requiring excessive computer time and storage. Usually the algorithm does not provide approximate solutions, only the exact solution on completion (if possible). Bellman[62] indicates where the approach is most likely to be appropriate in combinatorial problems.

14.10   INTEGER AND ZERO-ONE PROGRAMMING

When the standard programming problems are further restricted to variables with only these values, very difficult programming problems are defined. Although the problem formulation is concise, other 'in context' formulations are often preferable as the references show. At an elementary level, McMillan[3] summarises the integer linear theory mainly due to Gomory. Balinski[63] and Greenberg[64] give fuller treatments. Hadley[9], Hu[65], Abadie[42], Taha[66] and Garfinkel and Nemhauser[67] attempt to unify the theory, with the latter seeming most popular. Again McMillan[3] gives a good treatment of the Balas zero-one programming problems, although practical results have been discouraging. This topic seems as difficult as Diophantine analysis and as impractical (continued fractions help little with gear chains). Alternative combinatorial approaches seem preferable, and are mentioned in the above texts and discussed next.

14.11   TRANSPORTATION AND FLOW NETWORKS

Many operational research problems are in this field, which has an extensive theory and good computing algorithms. Engineering disciplines have natural networks and with new non-linear materials potential applications are numerous. Even when there is no obvious connection with networks, such a formulation often provides worthwhile insight.

The networks are abstract graphs with flows or potential differences associated with their arcs. As well as satisfying the appropriate Kirchhoff law, these are often limited to a finite interval when the network is termed 'capacitated'. The 'energy function', here usually cost, is to be minimised and only for convex functions is the theory complete. This implies a non-decreasing 'resistance law' resulting in a full theory of 'monotone capacitated networks'.

Minty[68] gave the key theorem and algorithm, but his abstract presentation precluded popularisation of the approach. In Minty[69] an analogy makes the graph colouring theorem obvious and more detail is given on the formulation aspects. Having understood this, the second half of Berge and Ghouila-Houri[40] displays the power of the ideas, many applications, and the vector space foundations. Problems of the shortest path, minimal spanning tree, maximal flow, and the maximal cost flow (transhipment and various transportation problems) are covered.

The earlier book by Ford and Fulkerson[70] covers similar material from a linear programming viewpoint, and despite its more restricted view is rich in theory and applications. Dennis[71] presents an even earlier view using electrical network analogies. An illuminating view of the maximum flow–minimum cut theorem is given with an analogy from plastic structures by Prager[72].

Non-convex and multi-commodity flow network problems are less complete theoretically and are surveyed by Fulkerson[73] and Hu[74]. Iri[75] gives a compact comprehensive study and Kaufmann[76] some unusual material.

At an introductory level, Busacker and Saaty[77] expound parts of the theory in a lively non-rigorous manner. Christofides[78] includes many algorithms in a more abstract setting.

14.12 COMBINATORIAL PROGRAMMING

This topic is flourishing to such an extent there is a department devoted to it at the University of Waterloo, Canada. Problems range from abstract graph colouring to practical machine-shop scheduling. The optimisation is over combinatorial entities such as combinations, permutations, partitions, finite graphs and partially ordered algebraic structures. A revival of interest in pure combinatorics has resulted in excellent new textbooks such as Berge[79].

Although the topic seems ideal for digital computation (viz. the hopes and quotations in Beckenbach[80]), the sheer size of the problems must be appreciated. With the fastest forseeable digital computers, complete enumeration of all schedules for every-day machine-shops would take 10^{10} years – the present age of the Universe! Mathematical theory is essential here to devise new approaches; hence the wide research interest. The large collection of papers cited by Roy[81] typifies the current situation. For mathematicians, the unaccustomed algorithmic viewpoint may require some acclimatisation.

Certain problems have become prominent, more for their testing of algorithmic ingenuity than for their practical value. Of these, Whitney's Traveling Salesman problem is typical, and has attracted much attention (see the survey by Bellmore and Nemhauser[82]). 'A salesman wishes to visit n towns and return to his starting town by the cheapest possible route. Given the costs of inter-town travel, not necessarily equal in opposite directions, determine the cheapest route'. For these $n + 1$ towns, there are $n!$ possible routes, one or more of which is the cheapest. Complete enumeration is limited to approximately 10 towns, and in this general formulation there is little current mathematical theory to help. For up to 18 towns dynamic programming is effective, but requires too much computer storage for larger problems. The 'Branch-and-Bound' procedure of Little *et al.*[83] is then surprisingly effective, and provides approximate solutions if terminated prematurely. Only from an algorithm viewpoint can the elegance of this versatile approach be understood. Even larger ($n > 100$) problems have been approached via Lin's[84] 'k-opt. algorithm', which again adapts to other difficult large problems (e.g. Kernighan and Lin[85]). This is now the realm of 'heuristics', using rules-of-thumb which are the only current techniques for large scheduling problems.

An avalanche of publications has begun in this field, mainly reported in the *Journal of Combinatorial Theory* (cf. Section 7.1). The books by Even[86], Wells[87] and Welsh[88] indicate the directions in which theory and applications are developing. As yet, there is little unity, although it seems likely that optimisation over abstract lattices, groups and other algebraic structures will provide this, to the interest of mathematicians. New fields such as 'computational complexity' are utilising this topic (see recent issues of the *Communications* and *Journal* of the Association for Computing Machinery).

REFERENCES

1. Kantorovitch, L., 'On the translocation of masses', *Akademiya Nauk SSSR. Doklady*, 37, 199-201 (1942)
2. Koopmans, T.C. (ed.), *Activity analysis of production and allocation*, Wiley, New York (1951)
3. McMillan, C., *Mathematical programming*, Wiley, New York (1970)
4. Abadie, J. (ed.), *Nonlinear programming*, North-Holland, Amsterdam (1967)
5. Fletcher, R. (ed.), *Optimisation*, Academic Press, New York (1969)
6. Dorn, W. 'Non-linear programming – a survey', *Management Science*, 9, 171-208 (1963)
7. Wilde, D.J. and Beightler, C.S., *Foundations of optimisation*, Prentice-Hall, Englewood Cliffs, N.J. (1967)
8. Hadley, G., *Linear programming*, Addison-Wesley, Reading, Mass. (1962)
9. Hadley, G., *Nonlinear and dynamic programming*, Addison-Wesley, Reading, Mass. (1964)
10. Zoutendijk, B. 'Non-linear programming: a numerical survey', *SIAM Journal on Control*, 4, 194-210 (1966)
11. Zangwill, W.I. *Nonlinear programming: a unified approach*, Prentice-Hall, Englewood Cliffs, N.J. (1969)
12. Mangasarian, O.L. *Nonlinear programming*, McGraw-Hill, New York (1969)
13. Saaty, T.L. and Bram, J., *Nonlinear mathematics*, McGraw-Hill, New York (1964)
14. Goldstein, A.A., *Constructive real analysis*, Harper and Row, New York (1967)
15. Geffrion, A.M., 'Elements of large scale mathematical programming, Part I: Concepts', *Management Science*, 16, 652-675 (1970)
16. Geffrion, A.M., 'Elements of large scale mathematical programming, Part II: Synthesis of algorithms and bibliography', *Management Science*, 16, 676-691 (1970)
17. Ackoff, R.L. and Sasieni, M.W., *Fundamentals of operations research*, Wiley, New York (1968)

18. Saaty, T.L., *Mathematical methods of operations research*, McGraw-Hill, New York (1959)
19. Vajda, S., *Readings in mathematical programming*, 2nd edn, Pitman, London (1962)
20. Mitchel, C.H., *Operational research*, English Universities Press, London (1972)
21. Nicholson, T.A.J., *Optimisation in industry*, 2 vols, Longman, London (1971)
22. Lavi, A. and Vogl, T.P. (eds), *Recent advances in optimisation techniques*, Wiley, New York (1966)
23. Riley, V. and Gass, S.I., *Linear programming and associated techniques*, John Hopkins Press, Baltimore, Md. (1958)
24. Kunzi, H.P. and Oettli, W., *Nichtlineare Optimierung: neuere verfahren Bibliographie*, Springer-Verlag, Berlin (1969)
25. Batchelor, J.H., *Operations research: an annotated bibliography*, St. Louis University Press, Missouri (1959)
26. Buckland, M.K., 'Sources of information for operational research studies', *Operational Research Quarterly*, 18, 297-313 (1967)
27. Kemeny, J.G., Snell, J.L. and Thompson, G.L., *Introduction to finite mathematics*, 2nd edn, Prentice-Hall, Englewood Cliffs, N.J. (1966)
28. Noble, B., *Applied linear algebra*, Prentice-Hall, Englewood Cliffs, N.J. (1969)
29. Trustrum, K., *Linear programming*, Routledge and Kegan Paul, London (1971)
30. Gass, S.I., *Linear programming: methods and applications*, McGraw-Hill, New York (1959)
31. Charnes, A., Cooper, W.W. and Henderson, A., *An introduction to linear programming*, Chapman and Hall, London (1953)
32. Dantzig, G.B., *Linear programming and extensions*, Princeton University Press, Princeton (1963)
33. Gale, D., *The theory of linear economic models*, McGraw-Hill, New York (1960)
34. Kuhn, H.W. and Tucker, A.W. (eds), *Linear inequalities and related systems*, Annals of Mathematics Studies 38, Princeton University Press, Princeton (1956)
35. Kuhn, H.W., 'The Hungarian method for the assignment problem', *Naval Research Logistics Quarterly*, 2, 83-97 (1955)
36. Flood, M.M., 'The traveling salesman problem', *Operations Research*, 4, 61-75 (1956)
37. Gale, D., 'Optimal assignments in an ordered set: an application of matroid theory', *Journal of Combinatorial Theory*, 4, 176-180 (1968)
38. Leontief, W.W., *The structure of the American economy 1919-1929*, Harvard University Press, Cambridge, Mass. (1941)
39. von Neumann, J. and Morgenstern, O., *Theory of games and economic behaviour*, Princeton University Press, Princeton (1944)
40. Berge, C. and Ghouila-Houri, A., *Programming, games and transportation networks*, Methuen, London (1965)
41. Rosen, J.B. *et al.* (eds.), *Nonlinear programming*, Academic Press, New York (1970)
42. Abadie, J. (ed.), *Integer and nonlinear programming*, North-Holland, Amsterdam (1970)
43. Boot, C.G., *Quadratic programming algorithms*, North-Holland, Amsterdam (1964)

44. Colville, A.R._A comparative study on nonlinear programming codes, Report No. 320-2949, IBM Scientific Centre, New York (1968)
45. Rosen, J.B., 'The gradient projection method for nonlinear programming, Part I: Linear constraints', *Society for Industrial and Applied Mathematics. Journal*, 8, 181-217 (1960)
46. Rosen, J.B., 'The gradient projection method for nonlinear programming, Part II: Nonlinear constraints', *Society for Industrial and Applied Mathematics. Journal*, 9, 514-532 (1961)
47. Goldfarb, D. and Lapidus, L., 'Conjugate gradient method for nonlinear programming problems with linear constraints', *Industrial and Engineering Chemistry. Fundamentals*, 7, 142-151 (1968)
48. Duffin, R.J., Zener, C. and Peterson, E.L., *Geometric programming*, Wiley, New York (1967)
49. Peterson, E.L., 'Geometric programming', *SIAM Review*, 18, 1-51 (1976)
50. Ortega, J.M. and Rheinbolt, W.C., *Iterative solution of nonlinear equations in several variables*, Academic Press, New York (1970)
51. Powell, M.J.D., 'A survey of numerical methods for unconstrained optimisation', *SIAM Review*, 12, 79-97 (1970)
52. Jacobi, S.L.S., Kowalik, J.S. and Pizzo, J.T., *Iterative methods for nonlinear optimisation problems*, Prentice-Hall, Englewood Cliffs, N.J. (1972)
53. Box, M.J.D., Davies, D. and Swann, W.H., *Nonlinear optimisation techniques*, Oliver and Boyd, Edinburgh (1969)
54. Box, M.J.D., 'A comparison of several current optimization methods, and the use of transformations in constrained problems', *Computer Journal*, 9, 67-77 (1966)
55. Fiacco, A.V. and McCormick, G., *Nonlinear programming: sequential unconstrained minimisation techniques*, Wiley, New York (1968)
56. Bellman, R., *Dynamic programming*, Princeton University Press, Princeton (1957)
57. Dreyfus, S.E., *Dynamic programming and the calculus of variations*, Academic Press, New York (1965)
58. Pontryagin, L.S. et al., *The mathematical theory of optimal processes*, Wiley, New York (1962)
59. Nemhauser, G.L., *Introduction to dynamic programming*, Wiley, New York (1966)
60. Bellman, R. and Dreyfus, S.E., *Applied dynamic programming*, Princeton University Press, Princeton (1962)
61. Kaufmann, A. and Cruon, R., *Dynamic programming*. Academic Press, New York (1967)
62. Bellman, R., 'Combinatorial processes and dynamic programming', *American Mathematical Society. Proceedings of symposia in applied mathematics*, 10, 217-249 (1960)
63. Balinski, M.L., *Approaches to integer programming*, North-Holland, Amsterdam (1975)
64. Greenberg, H., *Integer programming*, Academic Press, New York (1971)
65. Hu, T.C., *Integer programming and network flows*, Addison-Wesley, Reading, Mass. (1967)
66. Taha, H.A., *Integer programming: theory, applications and computations*, Academic Press, New York (1975)
67. Garfinkel, R.S. and Nemhauser, G.L., *Integer programming*, Wiley, New York (1972)
68. Minty, G.J., 'Monotone networks', *Royal Society of London. Proceedings, A*, 257, 194-212 (1960)

69. Minty, G.J., 'Solving steady-state nonlinear networks of "monotone" elements', *IRE Transactions on Circuit Theory*, CT-8, 99-104 (1961)
70. Ford, L.R. and Fulkerson, D.R., *Flows in networks*, Princeton University Press, Princeton (1962)
71. Dennis, J.B., *Mathematical programming and electrical networks*, Wiley, New York (1959)
72. Prager, W., 'Mathematical programming and theory of structures', *Society for Industrial and Applied Mathematics. Journal*, 13, 312-332 (1965)
73. Fulkerson, D.R., 'Flow networks and combinatorial operations research', *American Mathematical Monthly*, 73, 115-138 (1966)
74. Hu, T.C., 'Recent advances in network flows', *SIAM Review*, 10, 354-359 (1968)
75. Iri, M., *Network flow, transportation and scheduling*, Academic Press, New York (1969)
76. Kaufmann, A., *Graphs, dynamic programming and finite games*. Academic Press, New York (1967)
77. Busacker, R.G. and Saaty, T.L., *Finite graphs and networks*, McGraw-Hill, New York (1965)
78. Christofides, N., *Graph theory – an algorithmic approach*, Academic Press, New York (1975)
79. Berge, C., *Principles of combinatorics*, Academic Press, New York (1968)
80. Beckenbach, E.F., *Applied combinatorial mathematics*, Wiley, New York (1964)
81. Roy, B. (ed.), *Combinatorial programming: methods and applications*, Reidel, Dordrecht (1974)
82. Bellmore, M. and Nemhauser, G.L., 'The traveling salesman problem: a survey', *Operations Research*, 16, 538-558 (1968)
83. Little, J.D. et al., 'An algorithm for the traveling salesman problem', *Operations Research*, 11, 972-989 (1963)
84. Lin, S., 'Computer solutions of the traveling salesman problem', *Bell System Technical Journal*, 44, 2245-2269 (1965)
85. Kernighan, B.W. and Lin, S., 'An efficient heuristic procedure for partitioning graphs', *Bell System Technical Journal*, 49, 291-307 (1970)
86. Even, S., *Algorithmic combinatorics*, Macmillan, New York (1973)
87. Wells, M.B., *Elements of combinatorial computing*, Pergamon Press, Oxford (1971)
88. Welsh, D.J.A. (ed.), *Combinatorial mathematics and its applications*, Academic Press, New York (1971)

The following references are standard texts, but are not discussed in the chapter:

Beale, E.M.L., *Mathematical programming in practice*, Pitman, London (1968)
Collatz, L. and Wettering, W., *Optimierungs Aufgaben*, Springer-Verlag, Berlin (1966)
Graves, R.L. and Wolfe, O. (eds.), *Recent advances in mathematical programming*, McGraw-Hill, New York (1963)
Zoutendijk, G. *Methods of feasible directions*, American Elsevier, New York (1960)

Author Index

This index lists all authors cited in the book. The author approach to the literature is discussed in Section 1.2.

243

Subject Index

This index should be used in conjunction with the Contents List.
Journal titles are given in italics.